MW00845251

UNION IRON WORKS
SAN FRANCISCO, CAL.
STEAM
0 20 40 60 80 100 120 140 160 180 200 220 240

THE BOSS
14897
THE RUSSELL&COMPANY
BUILDERS
MASSILLON OHIO USA

Motorbooks International

FARM TRACTOR COLOR HISTORY

STEAM TRACTORS

HANS HALBERSTADT

DEDICATION

For several members of the ancient and honorable *Brotherhood of Threshermen*: Ellis Nelson, Lyle Hoffmaster and E. J. Murphy. And although he didn't do much thrashing, for John Bird, whose brilliant writing helped make *The Country Gentleman* a wonderful farm magazine half a century ago.

First published in 1996 by Motorbooks International Publishers & Wholesalers, 729 Prospect Avenue, PO Box 1, Osceola, WI 54020-0001 USA

Library of Congress Cataloging-in-Publication Data Available

ISBN 0-7603-0140-9

On the front cover: This marvelous Avery undermount steamer is one of the stars of the big annual show at Rollag, Minnesota. Like virtually every steam engine at every farm show, its preservation represents a labor of love—and a love of traditional American and Canadian family farm values. Restoring a steam engine to showroom condition like this requires a commitment that can last years and cost many thousands of dollars.

On the frontispiece: Instrumentation and controls may seem exotic and confusing to anyone accustomed to modern tractors or machinery, but steam engine builders provided engineers and firemen excellent systems for managing the output of their machines. Here are two essentials found on every steamer—the pressure gauge and the speed governor.

On the title page: Colorful Russell tractors are a much-beloved marque among steam traction engine fans. Russell & Company, like many who built steam traction engines for the agricultural market, started out making harvesting machines and later expanded into railroad and road-building heavy equipment. Russell tractors were built in Massillon, Ohio, until the 1920s.

On the back cover: (Top) Port Huron steam engines are among the most durable and respected examples of the breed. This one is getting ready to work at the wonderful little Hamilton, Missouri, show. (Bottom) With Terry Galloway, great-grandson of Daniel Best at the helm, this restored 110hp Best from the Oakland Museum spends time at Ardenwood Historic Farm near Fremont, California.

Printed in Hong Kong

CONTENTS

The steam powered fraternity of farm tractors prepares to greet the day at Rollag, Minnesota, on a Labor Day weekend. Some of the finest examples of the few surviving Buffalo Pitts steam tractors can be found here, hot to trot.

ACKNOWLEDGMENTS

Maybe it's no coincidence, but the older I get the more I enjoy and admire the wit and wisdom of old men. Not men of my own generation (which is old enough), but those of my father's and grandfather's day—old working men who remember life before The Big War, and old farming men in particular. They are in their late 70s and 80s today, and their boilers aren't putting out much steam pressure anymore, but they are worth collecting even so.

You meet them today at farm shows around the country, standing beside the tractor displays, leaning on canes and walkers, sometimes, or even in wheel chairs. The young punks among them—the guys in their 60s—admire the glittering green-and-yellow John Deere tractors, but the older, wiser men come to see the steamers.

I like listening to their stories about working from "can't-see-in-the-morning" to "can't-see-in-the-evening" to get a crop in the ground, or in before the storm, the praise of Avery tractors from some of them, or Russell, Case, or Advance Rumley from others.

The memories of hard work, of work well done, of people, times, places, events, successes, failures, the fabulous threshing-time dinners, stupid stunts, practical jokes, and brilliant repairs—all so many years ago. You don't see much of these men or their stories in history books, but they made history just the same.

These old men have a lot to say and they like to talk. I have a lot to learn and like to listen. This book is very much a product of what I have heard from these fine old men, and some fine young men, too. So my hat is off to:

Chuck Whitcher and Glen Christoffersen for the tractor anatomy lesson and the rare opportunity to view the dissected remains of their engines;

The Early Day Gas Engine & Tractor Association, and particularly my own Branch 3, for keeping the music alive;

Guy Fay, everybody's favorite bookworm and researcher;

Rick Halldorson, for all his help with the Avery 40 horsepower photos and research;

Robert Blades, and young Dane Blades too, who like to show their sturdy Port Huron tractor to a younger crowd;

Stan Mayberry, for generous helpings of stories, technical data, and for showing off at the Hamilton, Missouri, steam-up;

Lyle Hoffmaster, for all the great stories and insights;

Ellis Nelson, another fine, steam-powered gentleman with a full supply of wit and wisdom.

And I'd get booted right out of the American Brotherhood of Agricultural Authors if I didn't say something nice about Randy Leffingwell, a rare kind of friend who will loan photographs and reference materials to a brother author; thanks, buddy.

RUSSELL
THE
MASSILLON OHIO. U.S.A.
RUSSELL
17000
ENGINE

CHAPTER ONE

INTRODUCTION

Thrashing Day On the Farm

Off in the early-morning distance, down the lane that led from the county road, you could hear the muffled clatter and squeak of something coming to the farm. The arrival of the steam tractor and the threshing rig sometime around sunrise on a summer morning was the most anticipated, most delightful, most perfectly happy day of the year for millions of farm children in the days when our grandparents, or our great-grandparents were young. That was, in the years before World War II and before the self-propelled combine. It was before the gasoline tractor fully displaced the big, powerful, musical steam tractors that once roamed the heartland of America and Canada, to the delight of adults and children both. Boys and girls arose early and eagerly, made nests in the grass by the road and waited for the arrival of this exciting machine on this anticipated day—not willing to miss a moment of the experience, and intent on absorbing every thrilling sight and sound of this once-a-year event. The rattling and squeaking got louder as the procession approached, heard but not seen.

It's 6 A.M. and 50-plus iron monsters are waking up. The air is thick with coal and wood smoke and the sound of escaping steam.

OPPOSITE
This sturdy Russell is chuffing along in a businesslike way, maintaining an even 250rpm. On the other end of the belt is a sawmill slicing cedar logs into slabs of fresh lumber. Russell tractors were built at Massillon, Ohio, until the 1920s. They were and still are a favorite of many farmers, but that didn't save the company; Russell went out of business in 1927, killed off by the gasoline tractor.

Then, a series of musical toots from the whistle signaled the actual arrival of the steam tractor and the threshing outfit as the engineer turned off the county road and onto the lane leading to the farm yard. At last, around sunrise, the chuffing, puffing, clattering parade arrived. The dogs barked, the horses squealed, the small kids jumped up and down, and the adults watched with all the dignity they could manage. But it was difficult not to be impressed.

The steam traction engine was something clearly made of iron and steel, but something seemingly alive, too—it snorted and sighed and seemed to breathe. It had a heartbeat. Under load it hummed and roared purposefully, like a big black draft animal with huge wheels instead of feet. Its annual visit to the farm for threshing was just as wonderful as the arrival of the elephants that came to town with the circus, or maybe better.

"Thrashing" (as it was and is called in the Heartland) was a brief period in each farm's year, usually in July or August, when the small grain crop—barley, oats, wheat, mostly—was separated from the straw, cleaned, and made ready for market. On most farms, just one day of thrashing with a good steamer and separator, operated by a skilled crew, would clean the grain. Some big operations might need a few days or even a week, but not many.

J. I. Case

The sound of a steam traction engine is one element of its magic attraction. It looks like a machine, but it sounds like a living, breathing creature. It sighs, it has a heartbeat, it rumbles and roars and clatters and sometimes shrieks like a banshee. But this one is just puttering along so quietly you'd think all five tons of water, steel, and fire were no more than a big iron horse on tiptoe.

The whole year's effort: the plowing and planting and waiting and worrying; all the money for seed, equipment, the hired man, and the phosphate fertilizer—all came to a hard, sharp focus on that one day when the threshing machine and its crew came to "thrash" the crop. For farmers in Iowa, Oklahoma, Ohio, Montana, Nebraska, Saskatchewan, Alberta, and all the other places in North America where small grain is grown, thrashing day was "payday" for the whole year.

The steam traction engine made American and Canadian farm prosperity

Despite the apparent complexity, a steam engine is a very simple system. It is also a system capable of producing tremendous power at very low fuel cost. That single cylinder can easily produce far more horsepower than a conventional automobile engine, although published horsepower ratings might suggest otherwise.

By the 1930s manufacture of steam traction engines had stopped but thousands continued to help thresh the annual grain harvest right into the 1950s. This pickup doesn't really hold enough bundles to keep the thresher fed for long, but it was a lot faster getting out to the field and back with a load than a team and wagon.

This family farmer has set up right by the barn. The straw pile will be handy for bedding, an important by-product for operations with dairy cows or work horses.

possible. Without it the broad Great Plains, stretching from Texas up to the Canadian prairie, would have stayed an empty desert. The power of steam made large scale farming practical. The steam tractor broke the sod of the Great Plains and powered the separators and combines that harvested the grain crop. These machines didn't last very long. They appeared toward the end of the Civil War, in the late 1860s, only really became popular about 1890, and went out of production about 1925. They were big, heavy, expensive, complicated, dangerous, and required a very well-qualified engineer. They wore out rather quickly, within five or ten years typically. At the beginning of World War II most steam tractors were ready for retirement. By the beginning of World War II most American farmers wanted gasoline tractors, the latest thing. By the end of the war most surviving steam tractors were either completely worn out, cut up for scrap metal, or parked somewhere. Of the

The controls of most steam traction engines are simple, sturdy, and work about the same way. Throttle, clutch, petcock drain control, and a few other bells and whistles are all you need to know. Operating a steam engine safely, though, requires knowledge, judgment, and experience.

Bob Blades and his son Dane get the family's 1916 Port Huron ready for a day's work. The tractor came from central Iowa and was ready for work when it rolled off the low-boy. And work it does, powering a sawmill in northern Missouri, even today.

latter, most were left outside to the weather. Very, very few had their boilers and engines properly drained; a select few were parked in barns and machine sheds, prepared for storage, and protected.

But steam tractors were once so popular, well-constructed, and numerous that even today there are hundreds around the U.S. and Canada capable of getting up steam. You'll see them at farm shows, threshing bees, county fairs, plowing

Now here's a labor of love chugging down the furrow—Sonny Rowlands and Orman Rawlings' 1906 Advance 16hp tractor helping with the tillage chores at Corona Ranch, Temecula, California. When Sonny and Orman were resurrecting the Advance, the California Air Resources Board had a cow at the thought of somebody burning coal in such a machine. So here is perhaps the only LPG fueled steam tractor around; it makes 150psi just fine but you can't get a black cloud of this machine no matter what.

The annual grain harvest was still powered by steam when this photo was made in the late 1930s. A tender stands by to replenish the steamer around noontime on a summer day. The fireman has his hand on the hand pump lever used to transfer boiler feed water to the engine.

demonstrations and harvest festivals in nearly every state and province. There are about 60 steamers in residence at the big show at Mt. Pleasant, Iowa, every year, and many others roll in for the show. And 50 or more will be in operation at the really big show in Rollag, Minnesota, on Labor Day weekend. Over 100,000 people attend these shows, many primarily to see steam in action. You can't blame them, either, because it is just as thrilling today as it was 70 years ago, for those little kids watching the steam tractor and the "thrashing" rig come up the lane. In fact, some of those little kids of 1925 are the 80-year-olds of today—and they're still thrilled by the sight of steam.

LEFT
Young Dane Blades may only be 16 years old but he's a full-fledged steam traction engineer and one of the best in the business in northern Missouri. Except for the hat, Dane is much like many farm lads of 16 or 17 back around the time of World War I—boys who listened and learned and gained the respect of older men when they did a man's job in a man's way. Dane has his hand on the forward/reverse lever, waiting for a signal from the "separator boss."

The Marvelous Magic of Steam Power

Something really amazing happens when water is heated to the boiling point, a transformation that isn't appreciated today quite as well as 100 years ago. One researcher, a Professor P.S. Rose, described the power of steam in a 1910 book called *The Steam Engine Guide.* Prof. P.S. Rose starts with a conventional 25 horsepower steam engine boiler as an example. That boiler will hold 52 cubic feet-of water plus 26 cubic feet of steam at 150psi—pretty much plain vanilla specs for steam engines of the time, and of those shown in this book today. That

NEXT PAGE
Closed course! Professional driver! Do not try this in your own vehicle! Dane puts the pedal to the metal while dad pours on the coal. The Port Huron "red-lines" at 250rpm and does the standing-quarter-mile in about seven minutes.

Port Huron

This Keck-Gonnerman provides plenty of shade for the crew and for the boiler, too. It is wood fired here, but could use coal, straw, or oil with slight modifications.

26 cubic feet of steam pressurized at 150psi weighs just 9.73 pounds—but it holds 1,300,000 foot pounds of energy!

If this boiler were to explode (as they did quite often at the time), the result would come only partly from 1,300,000 foot pounds of steam pressure in the boiler. The 52 cubic feet of boiling water at that pressure would be heated to about 366 degrees Fahrenheit and would actually contain about 20 times the potential energy of the steam. If the pressure on that water is suddenly released, as happens when a boiler fails, the water literally explodes. The 3,000 pounds of water contains 38,000,000 foot pounds of energy to add to the energy in the steam. That's enough

LETTERS FROM *The American Thresherman and Farm Power*

My engine is a 22 horsepower Advance. While it is 7 years old, it has never had a leaky flue or stay bolt, and I have pulled almost everything with it, graders, stumps, eight mold-board plows, ten disk plows, and a house. It has a Canada high pressure boiler. We threshed 1,700 bushels of wheat and 400 bushels of oats on one "setting" with this rig, using $14 worth of coal. If any thresherman has threshed more grain at one "set" I'd like to hear about it. Such a run is unusual in Kay County.

C R Schmitt, Kildare, Oklahoma, in *The American Thresherman and Farm Power*, November 1922

No man can run any kind of a machine without expensive breakdowns, unless his engine is gone over every morning before starting. You must see that everything is in shape before the crew gets in or you may have some very embarrassing remarks made about your way of running machinery. Your trade will gradually drift away, and you will wonder why.

When it comes down to handiness and safety, the gas tractor is out of the race, for the steam engine cannot be beaten for any and all kinds of belt work. How many arms are broken each year in cranking tractors? How often must new bearings be put in a tractor on account of an oil feed being stopped? Did you ever see a tractor that would handle no load to full load without a noticeable change in speed? On the brake test, with a uniform load, the tractor will make a nice showing. For traction work, they are all right, but steam is good enough for me and my customers.

As to economy, I can show anyone, by figures and facts, that I can run my outfit, furnishing the coal, oil, separator man, water hauler, team, and engineer, cheaper per bushel of grain threshed than any gas tractor outfit furnishing fuel, oil separator man and engineer, that I have ever seen – and I have seen some of the best in this section of Ohio.

I own and operate the following machinery, besides running a 165 acre farm; one 23horsepower Baker engine with superheater, cab and bunks; one 33x56 Baker separator with Garden City feeder, Hart belt, and bucket weigher, and a Rockwood drive pulley, bought at our state fair in 1921; one No. 1 Birdsell clover huller with slat stacker; one No. 26 Appleton silo filler with 45ft of pipe; one No. 2 Standard Scheideler saw mill with 60in Ohlen inserted tooth saw; and a half interest in an Appleton No. 32D corn husker with shredder head, which surely made a record for me in 1921.

I notice some big days' work mentioned by other threshermen but I try to keep one gait, day in and day out. Do your hustling when getting ready to thresh, and then take a gait that you can keep. You will then find that at the end of the season you will have accomplished more work, and have less grain in the straw stack, and less repairs and breakdowns, than if you had tried to establish a new one-day record.

Ansel Wickliff, Pataskala, Ohio, in *The American Thershersman and Farm Power*, November 1922

Hi - ho, hi - ho, it's off to work we go . . . four Case steam engines ready for a ride to new owners. *J. I. Case*

Dane Blades at the helm of a Russell steam traction engine. Such stationary power work is standard fare for steam traction engines past and present.

power in that conventional, common, 25 horsepower boiler to fire a one-pound projectile straight up for 7,500 miles—into orbit. Professor Rose's point wasn't strictly for the sake of illustration; he and his readers all knew that you could do the math another way and shoot a 7,500 pound object (a steam tractor, for example) one mile straight up with the same energy.

But firing a steam tractor into low-earth orbit isn't a good use of all that energy. Instead, steam makes an excellent power source for doing work. Only within the last few years have the capabilities of steam locomotives been equaled by modern diesels. Back around the turn of the last century, 100 years ago, steam provided a practical way to get a lot of work done at an affordable price. Basically, that meant plowing and threshing. The steam tractor did both well. Sometimes, if you are lucky, you can still see one in action. A few folks still use them for farm work, even when nobody's watching, but hundreds are preserved and maintained and fired up to entertain crowds at farm shows around the U.S. and Canada.

RUSSELL
THE RUSSELL & CO.
MASSILLON OHIO, U.S.A.
RUSSELL
17000
ENGINE

ABOVE
A great many of America's roads were built with rigs like this one, a steam roller with tender and grader in tow, sometime about 1920.

TOP RIGHT
A lot of farmers got along just fine with a portable steam engine—the same power on the belt, but without the weight and complexity of a full steam traction engine. One of those portables is the star performer in this little drama, supported by a cast of 22 men and boys and four horses. *J. I. Case*

RIGHT
You won't find a lot of hired help these days that show up for work wearing a bowler hat, tie, and vest under their overalls but that's what our hero has done on the day this picture was made about 90 years ago. The engine is pulling a pair of 8-bottom disk plows, turning a 16ft swath. *J. I. Case*

CASE

440.

Many owners of steam traction engines used them for threshing during part of the year, roadbuilding during the rest. There are 14 lads on the crew, plus the usual contingent of "sidewalk superintendents" in this 1907 photograph. *J. I. Case*

TOP LEFT

Bagging grain at the separator is an old wild west custom seldom encountered east of the Rockies. There's a thousand bushels in that stack, more or less—a good morning's work. *J. I. Case*

LEFT

Eight huge steam-powered combines attack across the broad expanse of grain in this undated photograph. Such scenes were typical of the "bonanza" farms of the late 1800s and early 1900s—vast acreages, tremendous mechanization, and small numbers of hired hands. The bonanza farms lasted only a few years, till the soil gave out, then most were abandoned.

ILLINOIS
THRESHER
COMPANY

CHAPTER TWO

A HISTORY OF STEAM

Although James Watt gets the credit and the glory for developing the steam engine, he didn't do it alone, and he didn't do it first. Like most inventors, he pulled together a lot of earlier ideas into a new synthesis. The first steam engine we know of was an experimental technology demonstrator; the designer was an Egyptian named Hero and the engine was what we call today a "reaction steam turbine." That happened about 130 BC. The first *practical* steam engine was developed by Thomas Saver, and patented in 1698. It was a displacement vessel system that worked well enough to find wide use in pumping water from mines. It operated at between 100 and 150psi, didn't have a safety valve, and used fuel at a phenomenal rate. Saver's patent influenced another inventor, Denis Papin, who was working on a piston-in-cylinder design. His design, completed in 1705, was the most basic of steam engines, with the steam cylinder and the piston cylinder one basic component. The same year Thomas Newcomen modified these ideas with his own and demonstrated a walking-beam atmospheric engine. The valves for this engine had to be manipulated by hand—every cycle—and small boys were hired for the purpose.

The Avery Company, of Peoria, Illinois, built some of the most successful and innovative steam traction engines sold in Canada or the US. The company was founded in 1877 and, like nearly all other such companies, sold a full line of farm equipment: stationary, portable, and traction engines, plus threshing machines and "horse-power" systems. Although Case sold far more tractors, Avery billed itself as "the largest tractor company in the world." Avery was one of many casualties of the collapse of the farm economy that followed World War I, entering bankruptcy in 1924 and finally disappearing in the early 1940s.

OPPOSITE
This handsome, nicely restored steamer is performing at the Camp Creek show near Wahoo, Nebraska. Normally, picking up the grain bundles is a chore performed by the farmer's horses—always an essential member of the team during the days of steam—but the Camp Creek show has so many great steamers in attendance that these guys wanted to be outstanding in their field.

Now, according to conventional wisdom, James Watt came along in 1763 and invented the steam engine. Actually, he was busy fixing one of the Newcomen engines when he figured that a lot of useful heat and steam was being wasted. His contribution was the external condenser and a vacuum chamber. Both increased the efficiency and speed of the design. He also worked out a transmission system to convert the linear motion to rotary, complete with a flywheel to smooth out the whole process. About twenty years later, in 1782, Watt patented the "double-action" system and the "cut-off" that allowed the residual heat and pressure in exhaust steam to work a second piston.

By 1812 a 4 horsepower steam engine was powering a threshing machine. It reportedly threshed 1,500 sheaves in four hours. It's builder, Richard Trevithick, said at the time: "It is my opinion that every part of agriculture might be performed by steam. Carrying manure for the land, plowing, harrow-

The Best Manufacturing Co. was a major manufacturer of harvesting machines in the period following the Civil War and one of the earliest (1889) to design and build combines for grain harvesting. But Daniel Best's company's products didn't often travel outside California in those days when hundreds of small manufacturers served regional or local markets. This 1902 tractor was built at San Leandro, across the bay from San Francisco.

Even without that trademark you can usually spot a Buffalo Pitts tractor; no other manufacturer so frequently placed a round water tank right up front where it could block the engineer's view. The firm called Buffalo, New York, home for about 80 years, and was founded by twin brothers Hiram and John Pitts.

The first application for steam power on the farm was the use of stationary, rather than traction, engines to drive threshing machines and similar devices. This cut is from a 1885 German language catalogue promoting J. I. Case products, only seven years after the company began offering such engines. *J. I. Case*

ing, sawing, reaping, threshing, and grinding and all by the same machine, however large the estate. Even extensive commons might be tilled effectively, and without the use of cattle. I think a machine equal to the power of 100 horses would cost about £500. It would double the population of our Kingdom, for a great part of man's food now goes to horses which would then be dispensed with, and so prevent impression of corn, and at a trifling expense make our markets the cheapest in the world . . ." How right he was!

That same year, 1812, an American named Oliver Adams started doing some of what Trevithick predicted. Adams began building steam engines for farm use in his Delaware shop. Sixteen years later, in 1838, almost 2,000 stationary steam engines were working in the U.S., on and off the farm—far more than the number of railroad locomotives. These engines were of the low-pressure type, produced about 16 horsepower, and cost approximately $7,000 each.

By 1845 the cost was down to as little as $1,000 each and the numbers of farm steam engines was beginning to increase. But the design of these engines made them extremely stationary—bolted to solid masonry foundations in most cases, with separate boiler and engine assemblies—limiting their use. It was about this time that someone unknown combined the boiler with the engine in one compact assembly and put wheels underneath the whole business, creating the portable steam engine. This arrangement made the steam engine really practical for farm use.

This new mechanization came along just in time because the farms and cities of the U.S. (and a lot of foreign nations, too) were suddenly de-populated as millions of men (they were virtually all men, at first, except for a battalion or two of French prostitutes) dashed off to California in 1849. Until this time virtually all farm chores were done by hand; suddenly there were a lot fewer hands to help. American farmers were forced to mecha-

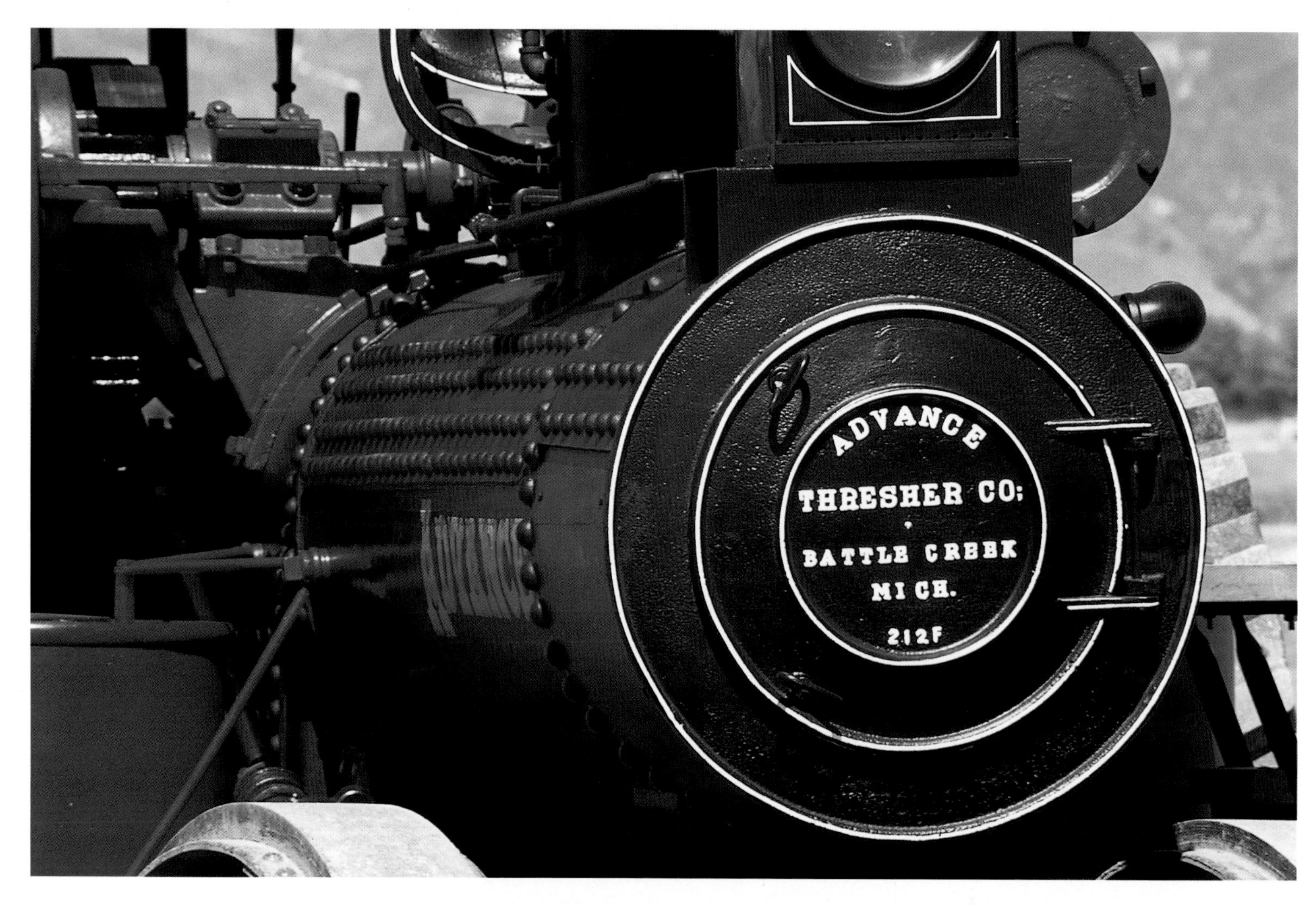

The many Advance steam traction engines you'll see frolicking at old time farm shows are a testament to the durability of this company from Battle Creek, Michigan. First known as the Advance Thresher Company and incorporated in 1881, the firm became part of a new company, the Advance Rumley Company in 1911. Advance, in its several permutations, built about 12,000 steam traction engines before production stopped in 1928. The firm became part of the huge Allis-Chalmers conglomerate in 1931.

A much-beloved marque among steam traction engine fans is the Massillon, Ohio, firm of Russell & Company. The company, like most of the others in the business of building steam traction engines for the agricultural market, started out making harvesting machines and later expanded into railroad and road-building heavy equipment. The company began business in 1842, incorporated in 1877, and went out of business in 1927.

An early steam traction engine from Case, an 1886 model. Unlike others available at the time, this example could be steered by the engineer; earlier versions needed horses for directional control. *J. I. Case*

nize to take up the slack. While the reaper, the mechanical thresher and separator, and the steam engine weren't developed in response to this new condition, they became popular partly as a result of the sudden lack of labor on the farm.

An early version, the "Forty-Niner," became available in 1849; this engine was built by A.L. Arcanbault in his Philadelphia, Pennsylvania, factory. Three sizes were available; you could buy a small 4 horsepower model for just $625 and the biggest 30 horsepower example was out the door for only $2,300 (taxes, license, transportation, and dealer prep was extra. Air conditioning wasn't an option). Until the Civil War these engines were strictly hand-crafted, home-brewed contraptions.

Other stay-at-home pioneers in the steam engine business were the Baker & Hamilton Company, a New York firm that started selling a steam engine for use with threshing machines in 1850; and the prolific Obed Hussey (who claimed to have the first reaper) with his steam plow engine in 1855.

President Abraham Lincoln promoted the idea of steam power on the farm in an address to the Wisconsin Agriculture Society in 1859, saying,

". . . The successful application of steam-power to farm work is a desideratum – especially a steam plow. It is not enough that a machine operated by steam will really plow. To be successful, it must, all things considered, plow better than can be done by animal power. It must do

Keck-Gonnerman was one of the few steam traction engine manufacturers to survive the transition to gasoline and kerosene engine technology.

all the work as well, and cheaper, or more rapidly, so as to get through more perfectly in season; or with some way, afford an advantage over plowing with animals, else it is no success."

The Civil War further disrupted the farm labor situation, improving the market for mechanization generally and for steam engines in particular. But practical steam tractors were still off in the distance. After the war steam engines became very common—but in the form of stationary or mobile power sources, not as self-propelled tractors. A 60 horsepower monster, the Standish Steam Plow, was plowing up to five acres an hour in 1868, about 50 times more ground in a day than a single plowman could typically work with a walking plow and a two-horse team. By 1900 about 5,000 steam tractors were being built each year, mostly by a handful of companies: J. I. Case, Huber, Advance-Rumley, Aultman & Taylor, and a few others.

Taking the Horse Out of Horsepower

Steam never came close to replacing the horse on the farm; that wasn't done until about 1940, and the gasoline tractor did the deed. In fact, steam power and horse power grew together over the years. About seven million horses were working on the farm in 1850, when the first steam farm engines appeared. When their numbers grew to 21 million in 1910, it was the apogee of steam power. That year four million steam horsepower and a half million gasoline horsepower engines were in use on the farm. Horses and steamers worked together on many farms, as a team. Horses and steam traction engines each had a set of virtues and vices that complemented each other. The numbers of both grew rapidly during the decades after the Civil War, ending around the time of World War I.

What happened? First, this was a

No glittering paint and chrome for John Tower's old Advance steamer. It was and still is a working machine on a working farm. Advance built about 12,000 such engines in their big factory at Battle Creek, Michigan, before the line shut down in 1928, years after most other companies quit the business.

long "Golden Age" for American farmers. The plains had been pretty well broken and homesteaded; an empty landscape filled up with people and farms, railroads, and towns. Wheat and small grain prices were, for much of this period, excellent. Steam technology had matured, evolved, and dropped in price. Steam power had proved itself to the American and Canadian farmer as a tremendous labor-saving device. But by about 1915, steam technology was evolved about as far as it could go on the farm—as a durable, powerful, economical source of power for stationary work like threshing, and draft work like plowing or road grading.

Boom and Bust on the Bonanza Farms

A whole new kind of farm was developed based on the steamer's tremendous power—the "bonanza" farms of Minnesota, the Dakota Territory, Montana, and even California. These farms specialized in small grains—mostly wheat, but also oats and barley—and were just huge. They could include an area under cultivation ten miles by ten miles—100 full sections! One such California farm had a wheat field 17 miles long! The bonanza farms were common for a brief while during the steam era, then the phenomenon disappeared.

Horses and mules had a role in these bonanza farms, but the steam tractor was the foundation of most. It was more economical, all around, for a big operation. The big steam tractors could—and did—operate around the clock, plowing with up to 50 bottoms, cutting swaths 30 feet across, mile after mile, to get the crop in before the weather closed down. It took capital, it took skilled labor, and it took money,

J. I. CASE & CO. RACINE WIS.

110 Horse-power Traction Engine

12 x 12-inch cylinder, simple

THIS is the largest and most powerful engine built in the CASE shops. It is designed for heavy plowing and freighting. As a Gold Medal Winner in each contest in which it has been entered at Winnipeg, it has demonstrated that there is no other steam engine made that is in its class. With its extra large capacity for carrying fuel —1,945 pounds of coal—and 366 gallons of water—nothing has been omitted for the convenience and efficiency of the operator. The pop valve is set for 160 pounds, but is allowed 170 pounds' pressure by the Canadian Provinces of Alberta and Saskatchewan.

This engine won the Grand Sweepstakes Prize as well as the Gold Medal at the Winnipeg Motor Contest, 1912.

Price complete, as shown above, including cab, $3,055.00. F. O. B. Racine, Wisconsin

Specifications

BOILER BARREL—38 inches in diameter.

FIRE-BOX—Length, 49¼ inches; width, 35¼ inches; height, 36 inches above grates. Stay Bolts, 1-inch diameter.

THROUGH-STAYS—Six steel through-stays, 1⅛-inch diameter, support the front and rear heads. Rear head has in addition four 1⅛-inch diagonal braces.

TUBES—76 in number, 2-inch diameter, 8 feet 4½ inches long.

HEATING SURFACE of boiler, 385 square feet (above grates).

GRATE AREA—12.06 square feet.

ROCKING GRATES are furnished regularly with this engine.

STEAM PRESSURE 160 pounds per square inch.

FLY-WHEEL—Diameter, 43½ inches; face, 16 inches; 230 revolutions per minute.

FRONT WHEELS—Height, 53 inches; tires, 14 inches wide regular or 20-inch special at extra price.

TRACTION WHEELS—Height, 7 feet; 36-inch tires. 12-inch extension rims will be furnished at extra price.

TRACTION SPEED—2.37 miles per hour.

EXTREME WIDTH of engine with 36-inch tires is 10 feet 8¼ inches; length, 21 feet 9 inches.

HEIGHT, to top of stack, 10 feet 5 inches.

DISTANCE between axles, 12 feet 2 inches.

WEIGHT, with boiler empty, 34,256 pounds.

Built regularly with simple cylinder engine and contractor's fuel bunkers, rocking grates and jacketed boiler.

SPECIAL ATTACHMENTS—Locomotive cab, extension rims, straw burner.

All CASE Engines will develop at least 10 per cent. more indicated horse-power than their actual guaranteed brake horse-power rating

Case 1913 catalogue description, 110hp steam traction engine. *J. I. Case*

LEFT
Here's Case's very first steam engine, an (approximately) 8hp portable built in 1869 and now installed in the Smithsonian. *J. I. Case*

but in virgin prairie soil, full of nutrients, the yields were terrific . . . at first. Operators of these big places ran them like industries, rather than farms in the old sense. They used their tractors until something broke—then discarded them and bought a new one. Many didn't last a year under these conditions.

Despite the tremendous cost of the operation, the bonanza farms were profitable during the first year. The second year's yield tended to be a bit lower—and you know what happened: the soil was quickly depleted, the farms used up the nutrients without putting anything back, and soon enough there wasn't much to sell. Profit margins hadn't been too wide to start with, but the economy of scale made it work. As the profit margin shrank, the bonanza farms went right out of business.

End Of the Line

But by this time gasoline engines had matured sufficiently as a technology that Americans were buying them in great numbers. An automobile frenzy developed in the years before World War I, followed by a gasoline tractor frenzy. All kinds of tractors, with all kinds of engines, traction systems, and specifications. Some were huge and mimicked the big steamers in size and shape and running gear. Others were small, light units that wouldn't break through anyone's bridge.

These gas tractors were much more convenient, safer, easier to operate, and the fuel was far more compact. No more shoveling coal, fussing about feed water, or watching the sight glass. You didn't have to get up two hours before you started work to get the boiler

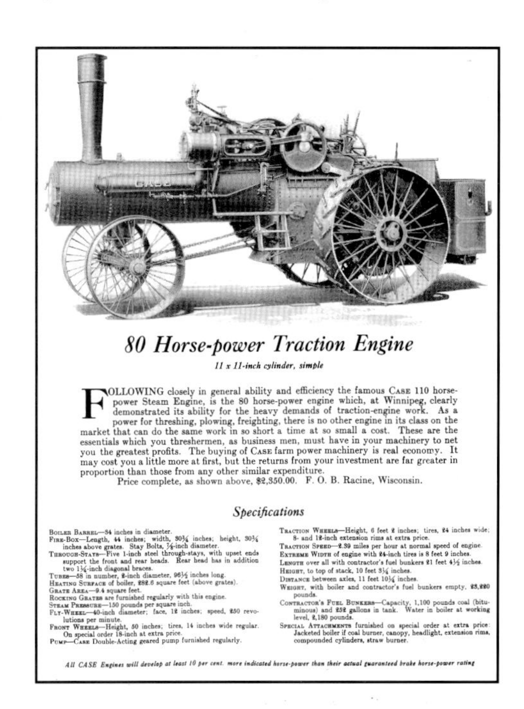

80 Horse-power Traction Engine

11 x 11-inch cylinder, simple

FOLLOWING closely in general ability and efficiency the famous CASE 110 horse-power Steam Engine, is the 80 horse-power engine which, at Winnipeg, clearly demonstrated its ability for the heavy demands of traction-engine work. As a power for threshing, plowing, freighting, there is no other engine in its class on the market that can do the same work in so short a time at so small a cost. These are the essentials which you threshermen, as business men, must have in your machinery to net you the greatest profits. The buying of CASE farm power machinery is real economy. It may cost you a little more at first, but the returns from your investment are far greater in proportion than those from any other similar expenditure.

Price complete, as shown above, $2,350.00. F. O. B. Racine, Wisconsin.

Specifications

BOILER BARREL—34 inches in diameter.

FIRE-BOX—Length, 44 inches; width, 30¾ inches; height, 30¾ inches above grates. Stay Bolts, ⅞-inch diameter.

THROUGH-STAYS—Five 1-inch steel through-stays, with upset ends support the front and rear heads. Rear head has in addition two 1¼-inch diagonal braces.

TUBES—58 in number, 2-inch diameter, 96½ inches long.

HEATING SURFACE of boiler, 282.6 square feet (above grates).

GRATE AREA—9.4 square feet.

ROCKING GRATES are furnished regularly with this engine.

STEAM PRESSURE—150 pounds per square inch.

FLY-WHEEL—40-inch diameter; face, 12 inches; speed, 250 revolutions per minute.

FRONT WHEELS—Height, 50 inches; tires, 14 inches wide regular. On special order 18-inch at extra price.

PUMP—CASE Double-Acting geared pump furnished regularly.

TRACTION WHEELS—Height, 6 feet 2 inches; tires, 24 inches wide; 8- and 12-inch extension rims at extra price.

TRACTION SPEED—2.39 miles per hour at normal speed of engine.

EXTREME WIDTH of engine with 24-inch tires is 8 feet 9 inches.

LENGTH over all with contractor's fuel bunkers 21 feet 4½ inches.

HEIGHT, to top of stack, 10 feet 3¼ inches.

DISTANCE between axles, 11 feet 10¼ inches.

WEIGHT, with boiler and contractor's fuel bunkers empty, 23,220 pounds.

CONTRACTOR'S FUEL BUNKERS—Capacity, 1,100 pounds coal (bituminous) and 252 gallons in tank. Water in boiler at working level, 2,180 pounds.

SPECIAL ATTACHMENTS furnished on special order at extra price: Jacketed boiler if coal burner, canopy, headlight, extension rims, compounded cylinders, straw burner.

All CASE Engines will develop at least 10 per cent. more indicated horse-power than their actual guaranteed brake horse-power rating

Case 1913 catalogue description, 80hp steam traction engine. *J. I. Case*

Case 1913 catalogue description, 75hp steam traction engine. *J. I. Case*

75 Horse-power Traction Engine

11 x 11-inch cylinder, simple

THIS engine drives our largest threshing machine, and works with equal certainty and ability in breaking virgin soil or in stubble land. William Vetleson, of Melford, Saskatchewan, has a 75 horse-power CASE engine and 36 x 58-inch Separator. "This equipment," he says, " is the best in the whole countryside for steady work, as it is never laid up for repairs."

Grismer Bros., of Grismerville, Saskatchewan, and Y. Y. Mosher, of Estevan, will attest to the ability of this engine.

It is from the users of machinery in the field that you can get the line on its real worth. Our booklet, "MONEY-MAKERS," gives a long list of names and addresses of CASE enthusiasts. We will gladly send it on request.

Price complete, as shown above, $2,150.00. F. O. B. Racine, Wisconsin.

Specifications

BOILER BARREL—34 inches in diameter.
FIRE-BOX—Length, 44 inches; width, 30¾ inches; height 30¾ inches above grates. Stay Bolts, ⅞-inch diameter.
THROUGH-STAYS—Five 1-inch steel through-stays, with upset ends support the front and rear heads. Rear head has in addition two 1¼-inch diagonal braces.
TUBES—58 in number, 2-inch diameter, 96½ inches long.
HEATING SURFACE of boiler, 282.6 square feet (above grates).
GRATE AREA—9.4 square feet.
STEAM PRESSURE—140 pounds per square inch.
FLY-WHEEL—40-inch diameter; face, 12 inches; speed, 250 revolutions per minute.
FRONT WHEELS—Height, 44 inches; tires, 12 inches wide regular. On special order 16-inch at extra price.
TRACTION WHEELS—Height, 5 feet 6 inches; tires, 24 inches wide; 8- or 12-inch extension rims at extra price.
TRACTION SPEED—2.5 miles per hour.
EXTREME WIDTH of engine with 24-inch tires is 9 feet 4¼ inches.
HEIGHT, to top of stack, 10 feet 2½ inches.
DISTANCE between axles, 11 feet 10¼ inches.
WEIGHT, with the boiler empty, and contractor's fuel bunkers, 20,440 pounds.
SPECIAL ATTACHMENTS furnished on special order at extra price: Jacketed boiler if coal burner, contractor's fuel bunkers, rocking grates, canopy, headlight, extension rims, compounded cylinders, straw burner.

All CASE Engines will develop at least 10 per cent. more indicated horse-power than their actual guaranteed brake horse-power rating

Case 1913 catalogue description, 60hp steam traction engine. *J. I. Case*

60 Horse-power Traction Engine

10 x 10-inch cylinder, simple

THIS engine, in more ways than any other, fills the needs of a general-purpose engine. Powerful for plowing, ideal for threshing, and eminently satisfactory in every way for freighting, hauling, sawmilling, pulling stumps, driving rock-crushers, husker-shredders, and a thousand and one uses, its ability is unrivalled. There is no engine that can compare with the 60 as a general-purpose engine.

Price complete, as shown above, $1,890.00. F. O. B. Racine, Wisconsin.

Specifications

BOILER BARREL—30½ inches in diameter.

FIRE-BOX—Length, 42 inches; width, 27 inches; height, 30½ inches above grates. Stay Bolts, ⅞-inch diameter.

THROUGH-STAYS—Four steel through-stays, 1-inch diameter, with upset ends, support front and rear heads. The rear head has in addition two 1⅛-inch diagonal braces.

TUBES—46 in number, 2-inch diameter, 90½ inches long.

HEATING SURFACE of boiler—216.9 square feet (above grates).

GRATE AREA—7.88 square feet.

STEAM PRESSURE—140 pounds per square inch.

FLY-WHEEL—40-inch diameter; face, 12 inches; speed, 250 revolutions per minute.

FRONT WHEELS—Height, 44 inches; tires, 10 inches wide regular.

TRACTION WHEELS—Height, 5 feet 6 inches; tires, 20 inches wide. 8- or 12-inch extension rims at extra price.

SPEED—2.61 miles per hour.

EXTREME WIDTH of engine with 20-inch tires is 8 feet 2 inches.

HEIGHT, to top of stack, 10 feet 1 inch.

DISTANCE between axles, 11 feet 2 inches.

WEIGHT, with boiler empty and contractor's fuel bunkers, 17,140 pounds.

SPECIAL ATTACHMENTS—Furnished on special order at extra price: Jacketed boiler if coal burner, contractor's fuel bunkers, canopy, headlight, extension rims, compounded cylinders, straw burner.

All CASE Engines will develop at least 10 per cent. more indicated horse-power than their actual guaranteed brake horse-power rating

The J. I. Case Company was the proverbial 800 pound gorilla of the steam traction engine industry, outselling the competition by about a 3 to 1 ratio and producing over 35,000 units of all kinds between 1878 and 1924. There isn't much sexy about any of them—they are all strong, sturdy, workhorses like this veteran getting up steam at Booneville, Missouri.

lighted; and you didn't have to worry about the ashes, the clinkers, or setting fire to the crop. Gasoline was cheap, simple fuel and gas tractors were cheap, simple machines by comparison. Once the reliability issue was resolved, it was no contest.

By the end of World War I, steam power was fading rapidly; by 1930 manufacture of steam tractors was over. But the steamers kept working, right through the Depression, through World War II, and even beyond. When they broke, they were parked in a field, or traded in on a gas tractor and cut up for scrap.

Ellis Nelson, an old-time Minnesota steam veteran, tells how they disappeared:

"This was in 1937. My dad had the old undermount Avery 30 horsepower and an old double cylinder Buffalo Pitts out in the weeds behind the blacksmith shop in West Hope, North Dakota. That Buffalo Pitts was a good engine, but it needed a couple of rivets replaced on one of the rear wheels, so we shipped it into town. Then we didn't have an immediate need for it, and it stayed there. It stood there in the weeds from 1916 until 1937. The scrap-pickers came along and found them – they chopped those engines up without even asking who owned them! They went to scrap iron for the Japanese – who shot them back at us! I didn't have the time or the money then to move those old engines someplace safe, and anybody who spent much effort fooling with steam engines in 1937 would have their sanity questioned, as well. But I think those engines would be worth a pile of money today, and I regret their loss every day."

Steam tractors still get used, occasionally, even today. John Tower uses his Advance to run a feed mill on his family's old Copperopolis, California, ranch, and Stan Mayberry powers his Dawn, Missouri, sawmill with his steam tractor (also an Advance) even today, when nobody's watching. But no serious routine farming is performed with them—so far as we know.

Steam Traction Engines–The Industry

Dozens of companies sold steam traction engines over the years, and you will still find many die-hard fans of long-dead companies like Port Huron, Peerless, Advance-Rumley, Russell, Reeves, Westinghouse, Wood Brothers, Huber, Buffalo-Pitts, and Keck-Gonnerman. There were many more. And some of those companies are still in business even if they aren't making steam traction engines. Here are thumbnail histories of some of the more important builders of steam traction engines.

Advance Thresher Co. and Advance-Rumley Co.

There are plenty of Advance engines still around, and still plenty of steam fiends with a preference for this brand. The company incorporated in 1881, primarily to build threshing machines. But Advance got into the steam traction engine business shortly thereafter. Over the fairly short period of 23 years, Advance cranked out over 12,000 steam traction engines of all kinds at its big factory at Battle Creek, Michigan, between 1888 and 1911.

Advance steam engines featured a short-stroke design based on a different understanding of the power output of an engine than was popular back before 1900. Then, many farmers and some engineers thought a long stroke utilized the potential of the steam pressure best; Advance designers knew the real key to efficiency was *piston travel* during a given period, not revolutions.

Advance was bought out by Rumley in 1911 and became part of a little conglomerate called Rumley Products. That latter company was reorganized in 1915 and the fabled Advance-Rumley Company was one of the spin-offs. Advance-Rumley stayed in business until 1931, absorbing another steam manufacturer, Autlman–Taylor, in 1915. Advance-Rumley made steamers later than most other builders, right up to 1928. The company's assets were absorbed in yet another acquisition, this time by Allis–Chalmers Corporation, in June of 1931.

C. Aultman Co. and Aultman – Taylor Machinery Co.

Cornelius Aultman was one of many skilled craftsmen of the middle 1800s who designed and built tools and equipment for the farm trade. He set up his company in 1851 and began building steam traction engines during the latter part of the century. The factory was located in Mansfield, Ohio.

The Aultman Co., and then the Aultman–Taylor Machinery Co., built a great number of steamers. The roster shows almost 6,000 units until they shut the line down in 1924. Most used high pressure engine systems and a 150psi working pressure. Some used a "water bottom" boiler that fully enclosed the firebox for increased heating surface area (and complexity).

50 Horse-power Traction Engine

9 x 10-inch cylinder, simple

THIS 50 horse-power engine, like all CASE engines, is a great worker on the highway. It is sure to go through all sorts of roads, and there are no hills within reason that it cannot climb. To co-operative threshermen this element is as important as its ability to drive a threshing machine. Figure your losses in case you should be "stuck" in going from one job to another. Your crew is idle, your machine is idle. Idleness in an investment is waste, and waste in this case is loss of dollars and cents. These are the things you business-threshermen must consider. If you investigate closely you will come to the conclusion that for real economy you must buy a CASE engine.

Price complete, as shown above, $1,670.00. F. O. B. Racine, Wisconsin.

Specifications

BOILER BARREL—28 inches in diameter.

FIRE-BOX—Length, 39½ inches; width, 25¼ inches; height, 30 inches above grates. Stay Bolts, ⅞-inch diameter.

THROUGH-STAYS—Four steel through-stays, ⅞-inch diameter, with upset ends, support front and rear heads; rear head has in addition two 1⅛-inch diagonal stays.

TUBES—36 in number, 2-inch diameter, 84½ inches long.

HEATING SURFACE of boiler, 164.7 square feet (above grates).

GRATE AREA—6.9 square feet.

STEAM PRESSURE—150 pounds per square inch.

FLY-WHEEL—40-inch diameter; face, 12 inches; speed, 250 revolutions per minute.

FRONT WHEELS—Height, 44 inches; tires, 10 inches wide regular.

TRACTION WHEELS—Height, 5 feet 6 inches; tires, 20 inches wide; 8- or 12-inch extension rims at extra price.

SPEED—2.3 miles per hour.

EXTREME WIDTH of engine with 20-inch tires is 7 feet 6¼ inches.

HEIGHT, to top of stack, 9 feet 11 inches.

DISTANCE between axles, 10 feet 6 inches.

WEIGHT, with the boiler empty and contractor's fuel bunkers, 16,480 pounds (estimated).

EXTRA ATTACHMENTS furnished on special order at extra price: Jacketed boiler if coal burner, contractor's fuel bunkers, canopy, headlight, extension rims, compounded cylinders, straw burner.

All CASE Engines will develop at least 10 per cent. more indicated horse-power than their actual guaranteed brake horse-power rating

Case 1913 catalogue description, 50hp steam traction engine. *J. I. Case*

40 Horse-power Traction Engine

8¼ x 10-inch cylinder, simple

IN the smaller engine classes there is no machine built that can equal in efficiency this 40-horse-power Traction Engine. As a powerful, easily handled, well-adapted engine for general tractioning it stands without an equal. In the 1912 Winnipeg Motor Contest this engine developed, in actual test, 63.35 horse-power—more than substantiating our claims that all CASE engines will develop at least 10 per cent. more indicated horse-power than their actual guaranteed brake horse-power rating.

For reference to its ability, we refer you to any user. They all will attest to their experiences with this or other CASE machinery, whether it be used for hauling, threshing, or any other of the uses to which CASE engines can be put.

Price complete, as shown above, $1,480.00. F. O. B. Racine, Wisconsin.

Specifications

BOILER BARREL—26 inches in diameter.

FIRE-BOX—Length, 35 inches; width, 25¼ inches; height, 30 inches above grates. Stay Bolts, ⅞-inch diameter.

THROUGH-STAYS—Four steel through-stays, ⅞-inch diameter, with upset ends, support front and rear heads; rear head has in addition two 1⅛-inch diagonal stays.

TUBES—30 in number, 2-inch diameter, 77 inches long.

HEATING SURFACE of boiler, 130.2 square feet (above grates).

GRATE AREA—6.14 square feet.

STEAM PRESSURE—150 pounds per square inch.

FLY-WHEEL—40-inch diameter; face, 10½ inches; speed, 250 revolutions per minute.

FRONT WHEELS—Height, 44 inches; tires, 10 inches wide regular.

TRACTION WHEELS—Height, 5 feet 6 inches; tires 18 inches wide; 8- or 12-inch extension rims at extra price.

SPEED—2.35 miles per hour.

WIDTH of engine with 18-inch tires is 7 feet 4¾ inches.

HEIGHT, to top of stack, 9 feet 9½ inches.

DISTANCE between axles, 9 feet 5¼ inches.

WEIGHT, with boiler empty and contractor's fuel bunkers, 14,810 pounds.

ATTACHMENTS furnished on special order at extra price: Contractor's fuel bunkers, jacketed boiler if coal burner, canopy, headlight, extension rims, compounded cylinders and straw burner.

All CASE Engines will develop at least 10 per cent. more indicated horse-power than their actual guaranteed brake horse-power rating

LEFT
Case 1913 catalogue description, 40hp steam traction engine. *J. I. Case*

Unlike some other companies, Aultman was ecumenical about technology—they'd build and sell just about anything: direct flue, return flue boilers; simple engines, compound engines, overmounted and undermounted, with single cylinders and dual cylinders, burning straw, coal, wood, and maybe a few that were nuclear powered. One of the odd features of some Aultman steam traction engines is a somewhat unconventional output shaft from the engine and a compensating geartrain to the drive wheels.

Avery Co.

The Avery Company built some of the most handsome, well-proportioned and powerful steam traction engines in the whole history of the industry. The company was founded in 1877 in Galesburg, Illinois, as a designer and manufacturer of farm tillage implements and planters. The company moved to Peoria, Illinois in 1884 and added steam traction engines and threshers to the product catalogue in 1891.

Steam engines and threshers became the company's stock in trade until the end of the line, thirty years later. The company billed itself for a while as "the largest tractor company in the world," despite the size and output of the J.I. Case Co., normally far greater. But Case was a full-line company, not a tractor specialist, and perhaps that is how Avery justified the claim.

The glory days for the company were the years before World War I, the years when the Avery Undermount line was one of the most capable, sophisticated, durable tractors in the field.

Although the 1920s are considered a prosperous prelude to the depression of the 1930s, the 1920s were tough on the farm, the farmer, and the whole economy that served rural America. After many years of high grain prices, the bottom fell out of the market after

Reeves steam traction engines are not nearly as common today as the products of Case or Advance, but they were well liked and respected in the past and remain so today. The Reeves Company started business after the Civil War, incorporated in 1888, and built steam engines for all kinds of farm uses. Emerson-Brantingham bought up Reeves in 1912 but continued to sell Reeves products until 1925 when the plant was closed.

80 Horse-power Portable Engine

11 x 11-inch cylinder, simple

OUR line of Portable Engines, similar to the ones illustrated, comprises the 30, 40, 50, 60 and 80 horse-power sizes. They are built along the same lines as our traction engines, have the same type of boiler, the same fittings and fixtures. The rear end is supported by extra strong stub axles, which are held by substantial brackets bolted to the side sheets of the boiler. The large wheels, of our regular steel rim and spoke type, with long bearings in the hubs and smooth tires, insure light draft.

All sizes can be fitted with our straw-burning attachment, except the 18 and the 30, including jacketed boiler, at an extra cost of $50. Brakes for all sizes are furnished, but only on special order at an extra price.

Price complete, as shown above, $1,150.00. F. O. B. Racine, Wisconsin.

Specifications

BOILER BARREL—34 inches in diameter.

FIRE-BOX—Length 44 inches; width, 30¾ inches; height, 30¾ inches above grates. Stay Bolts, ⅞-inch diameter.

THROUGH-STAYS—Five 1-inch steel through-stays, with upset ends support the front and rear heads. Rear head has in addition two 1¼-inch diagonal braces.

TUBES—58 in number, 2-inch diameter, 96½ inches long.

HEATING SURFACE of boiler, 282.6 square feet (above grates).

GRATE AREA—9.4 square feet.

STEAM PRESSURE—150 pounds per square inch.

FLY-WHEEL—40-inch diameter; face, 12 inches; speed, 250 revolutions per minute.

FRONT WHEELS—Height, 42 inches; tires, 10 inches wide.

REAR WHEELS—Height, 53 inches; tires, 12 inches wide.

EXTREME WIDTH of engine is 7 feet 5½ inches.

HEIGHT, to top of stack, 10 feet 2½ inches.

DISTANCE between axles, 10 feet 1 inch.

WEIGHT, with the boiler empty, 12,280 pounds; fitted regularly with reversing valve gear and CASE geared pump.

SPECIAL ATTACHMENTS furnished on special order at extra price: Jacketed boiler if coal burner, rocking grates, headlight, brake, and straw burning attachment.
Built with simple or compounded cylinders. Furnished as coal and wood, or straw burner.

All CASE Engines will develop at least 10 per cent. more indicated horse-power than their actual guaranteed brake horse-power rating

30 Horse-power Traction Engine

7¼ x 10-inch cylinder, simple

THIS is the smallest traction engine built in the CASE Company's shops. It combines in a remarkable degree, but on a smaller scale, of course, the efficient qualities of the 110, the 80, the 75, the 60, the 50, and the 40. It is built to drive small threshing machines, corn-shellers, feed-grinders, rock-crushers, cotton-gins; to haul logs or to do anything else for which a small engine can possibly be used.

Price complete, as shown above, $1,225.00. F. O. B. Racine, Wisconsin.

Specifications

BOILER BARREL—26 inches in diameter.

FIRE-BOX—Length, 30 inches; width, 23¼ inches; height, 23⅛ inches above grates. Stay Bolts, ⅞-inch diameter.

THROUGH-STAYS—Four steel through-stays, ⅞-inch diameter, with upset ends, support front and rear heads; rear head has in addition two ⅝-inch x 2½-inch flat diagonal braces.

TUBES—38 in number, 1¾-inch diameter, 67 inches long.

HEATING SURFACE of boiler, 116.7 square feet (above grates).

GRATE AREA—4.85 square feet.

STEAM PRESSURE—140 pounds per square inch.

FLY-WHEEL—36-inch diameter; face, 9½ inches; speed, 250 revolutions per minute.

FRONT WHEELS—Height, 38 inches; tires, 8 inches wide.

TRACTION WHEELS—Height, 4 feet 5 inches; tires, 14 inches wide regular; 8-inch extension rims at extra price.

TRACTION SPEED—2.26 miles per hour.

EXTREME WIDTH of engine with 14-inch tires is 7 feet 1 inch.

HEIGHT, to top of stack, 8 feet 9½ inches.

DISTANCE between axles, 8 feet 1½ inches.

WEIGHT, with boiler empty, including regular tank, 11,253 pounds.

ATTACHMENTS on special order at extra price: Jacketed boiler, canopy, headlight, contractor's fuel bunkers, extension rims.

All CASE Engines will develop at least 10 per cent. more indicated horse-power than their actual guaranteed brake horse-power rating

Case 1913 catalogue description, 30hp steam traction engine. *J. I. Case*

LEFT
Case 1913 catalogue description, 80hp steam traction engine. *J. I. Case*

World War I. Farmers scrambled to stay in business by avoiding the purchase of new equipment (like tractors and threshers) or by defaulting on the loans for the equipment they bought. Avery, like Case, Deere, and most other established manufacturers, had very liberal credit policies. Farmers paid their debts when there was money to pay them with; after the War To End All Wars, many farmers were broke even when the harvest was in and sold. Avery went bankrupt in 1924. It stayed barely afloat for the next few years, making some money in the early 1930s, then disappearing in the 1940s—but not without a trace. Avery remains one of the emotional favorites of many "steam fiends" and Avery tractors are likely to be chugging around farm shows for another century or two, at the least.

Avery Undermount Tractors, 1914 Models Compared

Avery offered seven variations on the undermount theme for the 1914 catalog, all with the same basic shape and layout, differing primarily in size. The smallest, rated at 18 horsepower, closely resembled the largest, at 40 horsepower, and all within the line shared many tried-and-true design features.

The company (along with most of the industry at this time) had settled on the straight flue, locomotive-type boiler. Unlike the rest of the industry, Avery used two cylinders for smoother power and the undermount layout for convenient engine maintenance. The boiler itself was from 32 inches to 36 inches in internal diameter, with 2-inch or 2.5-inch flues, and both the smallest 18 horsepower and the largest 40 horsepower boilers were 84 inches long. The number of flues varied between 41 and 58. The boiler shell for all but the smallest was 3/8 inch, typical for the industry at the time. These configurations provided a heating surface of 216 square feet (18 horsepower) to 330 square feet (40 horsepower). Working pressure for these boilers varied from 150 to 200psi.

Engines on these steam traction engines were also quite standardized, each with a 10-inch stroke and a 6-inch or 7-inch bore. In fact, apparently just two engines were used, a 6x10 for the 18, 20, and 20 Special models, and a 7x10 for the 22, 30, 30 Special, and 40 horsepower models. The variation in power output seems to have been determined by the

NEXT PAGES
Advance steam traction engines were among the most popular brands on the market back before World War I. This 1905 example is probably a lot cleaner and prettier now than it ever was, even on the showroom floor over 90 years ago. Advance built around 12,000 steam traction engines in 12 sizes, from 10hp to 40hp.

DVANCE THRESHING MACHINERY
ADVANCE

Peerless steam traction engines were built by the Geiser Manufacturing Company at Waynesboro, Pennsylvania, and this one is still going strong at the Adrian, Missouri, annual show. Geiser's steam engines and the rest of the product line became part of Emerson-Brantingham in 1912. That's a portable steam engine in the background, not a steam traction engine.

pressure rating of the boiler; all were rated at 250rpm.

The drivetrain for the Avery models of 1914 were designed to handle the strain of heavy plowing—3-inch steel shafts for the intermediate power output, up to 6 inches of solid steel for the axle shaft of the larger designs. Rear drive wheels were from 65 inches to 80 inches diameter. Two road speeds were provided on all models, about 1.9 miles per hour at 250rpm in low, about 2.6 miles per hour in high gear. Water capacity for the models varied from 325 gallons to 450 gallons, and coal capacity ranged from 400 to 500 pounds.

Best Manufacturing Co.

Daniel Best found gold in California, but he didn't have to dig it from the ground. A huge agricultural boom occurred in the far west just after the Civil War, with "bonanza" farms and tremendous acreage. California was the nation's leading wheat producer in 1880, and that production was done with a minimum of hand labor. Best and others of the time designed and built innovative farm machines for this trade, including a functional combine way back in 1889.

Part of that combine was Best's prototype steam traction engine, a monster weighing 22,000 pounds. This steamer, and the rest of his line of machines, were built in sunny San Leandro, on the eastern shore of San Francisco Bay. Best steam traction engines used a vertical boiler system that offered some advantages and some real headaches. Nearly all Best tractors were quite tall, often over 17 feet, limiting navigation on most farms. You couldn't drive your Best tractor into many unmodified barns!

The Best Manufacturing Co. engaged in a long series of court battles with the Holt Company, a competitor from nearby Stockton, California. It was finally resolved in 1925 by a merger of the two companies in a union that is now the Caterpillar Tractor Co.

The assembly line, before robotics. These steam traction engines are being assembled in place; units are in all phases of completion in this circa 1900 photo. *J. I. Case*

Profile: A Case History

The J. I. Case Company dominates much of American farm history, particularly the section on steam traction. Case, essentially, blew the doors off the competition, outselling all other builders by about a three-to-one ratio between 1880 and 1924, producing over 35,000 steam engines (portable and traction) during that time.

The Case company is one of the oldest in continuous operation in the U.S., almost as old as John Deere, and both are still going strong after 150 years of serving the American farmer. Case started in 1844 with a line of simple "groundhog" threshing machines. Through very astute management and marketing, Case was soon a major player in the new farm

18 Horse-power Portable Engine

6 x 8-inch cylinder, simple

THIS is the smallest portable engine we make. It is built with the same thorough attention and care as the larger ones. Its purpose is to run small threshing machines, wood-saws, grinders, corn-shellers, hay-balers and pumps, and for general light stationary work. This engine with an 18 x 36-inch CASE Threshing Machine makes an excellent outfit for all owners of medium-sized farms.

"My Case Portable Engine is the best engine I have ever used. I use it for sawing timber of all kinds and fire with green slabs right off the log, and it has all the power I possibly need."
Dana, N. C., July 31, 1912. FURMAN JONS

Specifications

BOILER BARREL—22 inches in diameter.

FIRE-BOX—Length, 26 inches; width, 20 inches; height, 25¼ inches above grates. Stay Bolts, ⅞-inch diameter.

THROUGH-STAYS—One steel through-stay, 1-inch diameter, supports front and rear heads. Rear head has in addition two flat diagonal braces, ½ inch x 2¼ inches.

TUBES—30 in number, 1¾-inch diameter, 54 inches long.

HEATING SURFACE of boiler, 80.9 square feet.

GRATE AREA—3.6 square feet.

STEAM PRESSURE—140 pounds per square inch.

FLY-WHEEL—36-inch diameter; face, 7 inches; speed 250 revolutions per minute.

FRONT WHEELS—30-inch diameter; tires, 5 inches wide.

REAR WHEELS—36-inch diameter; tires, 5 inches wide.

WEIGHT, with boiler empty, 4,686 pounds.

HEIGHT, to top of stack, 7 feet 8½ inches.

THE BRAKE has sufficient power to control the engine on steep grades. It is included only when ordered at extra price of $10.
Feed-water heater and steam pump are omitted. Two injectors are furnished.
Built only with simple cylinder as a portable engine. Coal and wood burner.

All CASE Engines will develop at least 10 per cent. more indicated horse-power than their actual guaranteed brake horse-power rating

Case Boilers. Case 1913 catalogue description. *J. I. Case*

LEFT
Case 1913 catalogue description, 18hp steam portable engine. *J. I. Case*

Case Boilers

We make our boilers from the best grade of open-hearth flange steel, having a tensile strength of 60,000 pounds per square inch

NO motor power has yet been devised to take the place of steam for threshing, plowing, hauling, grading, or any other demand for traction power. There are some localities where coal and water are at a premium, and there the CASE Gas and Oil Tractors meet all demands. For general use, however, CASE Steam Engines still maintain their supremacy as the leading traction power.

Boilers for traction-engine work must be made of the very best materials and particularly strong, because the traction engine, going up and down hill, over the rough as well as smooth roads, is severely strained from every angle. These strains must be met in its construction. The service of a traction-engine boiler is very different from that of a stationary engine, or even that of the boiler of a locomotive. The former rests on a foundation as solid as rock, and the latter is mounted on wheels which run on tracks, giving the boiler very little strain. We make all CASE Boilers in our own shops, and in their design and manufacture every jolt and bump that they may receive on the road and in the field is carefully provided for. All material that goes into these boilers is submitted to physical and chemical tests in our own laboratories. Our system, that compels all material bought to conform to our specifications and an analysis made in our own laboratory after delivery, makes it impossible for a defective part to get into a CASE Boiler.

We have the largest and best equipped boiler plant in the world devoted exclusively to the building of boilers for traction and portable engines. All boilers are tested by both hydraulic and steam pressure before leaving the factory.

Shop Tests. Boilers up to the 110 horse-power size are tested at a hydraulic pressure of 225 pounds; the 110 horse-power at a pressure of 260 pounds. This test is made in the boiler shops. After the completed engine leaves the erecting department, it is kept under steam for several hours and worked on a Prony Brake, developing fully its advertised horse-power. All boilers are kept under a pressure of 140 pounds of steam, excepting the 110 horse-power size, which is tested at 160 pounds. The Government inspectors of the provinces of Alberta and Saskatchewan, Canada, allow them a steam pressure ranging from 146 to 175 pounds per square inch, which allows a liberal margin above the required pressures.

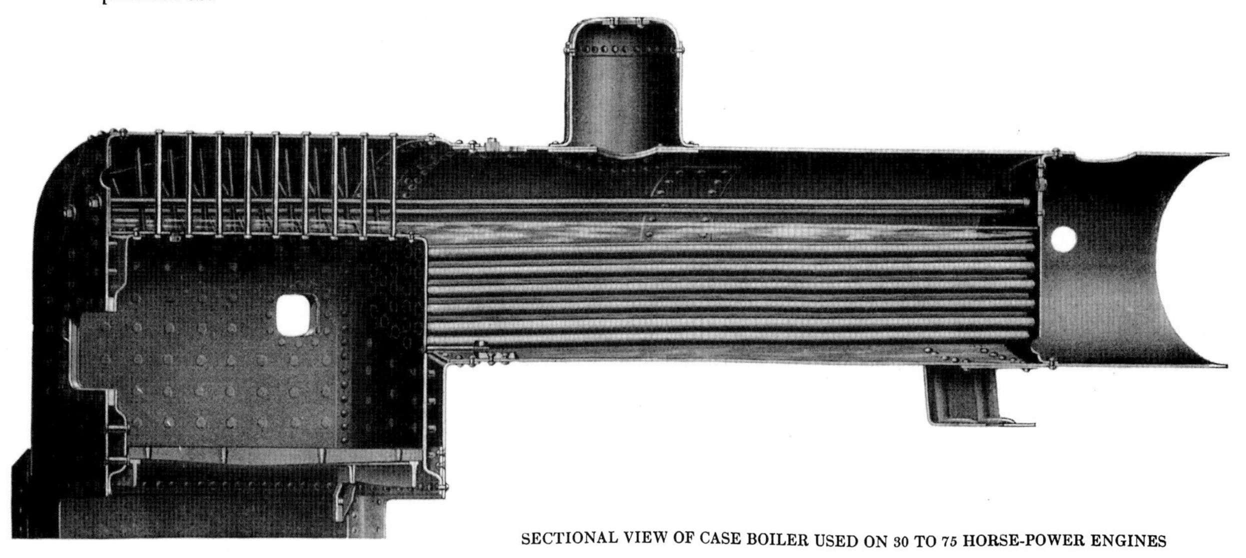

SECTIONAL VIEW OF CASE BOILER USED ON 30 TO 75 HORSE-POWER ENGINES

CASE
J.I.CASE THRESHING MACHINE.CO.
RACINE. WIS. U.S.A.

Oblique view, Case 80hp Traction Engine. Case 1913 catalogue description. *J. I. Case*

machine industry that developed about the time of the California Gold Rush of 1849. While Deere concentrated on plows and tillage tools, Case built harvesting machines—"separators," as the early threshing machines were called, and related products.

Case expanded its line during those early years. By 1876 the company was offering its own steam engines, primarily to power its own (and anyone else's) threshing machines and feed mills. Those first engines were strictly the portable power source that were common at the time, moved from place to place behind a team of horses. These first engines were of 8 horsepower and 10 horsepower size; 75 were built during that first production year.

Stationary and portable steam engines were part of the Case line for many years, right up until the bitter end. The last Case steamer built in 1924 was a portable, not a tractor, quite like the first farm engines of 1850.

But the market for a self-propelled steam traction engine—the "tractor," as it came to be called—was powerful enough, and well developed enough (by other manufacturers) that Case started providing running gear on some models, beginning in 1878. These first tractors used gear and shaft drives for motive power but still required horses for steering, a system that Case used until 1884.

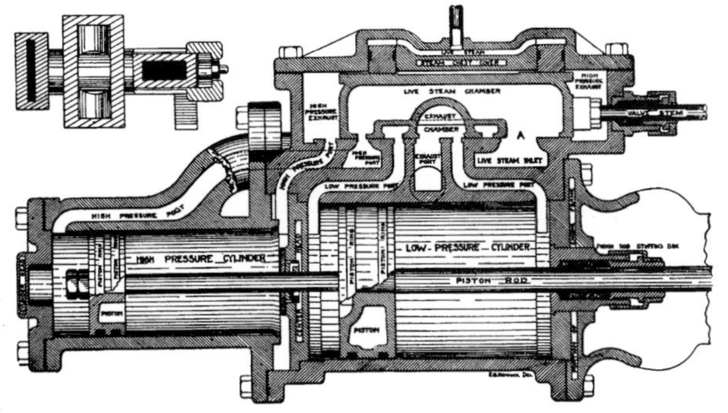

The market for steam farm engines, tractors or portables, was essentially flat during the last quarter of the nineteenth century. Case sold around 200 to 300 units a year during this period, right up until 1898 when only 211 were built. But then the market took off, starting the glory days of the steam tractor epoch, from 1899 until 1915. Production more than quadrupled for Case in just one year, up to 920 units in 1899, then over 1,000 in 1900, and running at about 1,500 to 2,000 engines each year for the next decade. Over 2,300 shipped in 1911 in sizes from 18 horsepower to giant 110 horsepower models.

Case sold 2,252 steam engines in 1912, and it was all down-hill from there. Two years later only about half that number sold, and by 1917 the number was down to just 598. American farmers were buying tractors like crazy—gasoline, kerosene, and distillate tractors without boilers, without fussy requirements for fuel and water, without the possibility of blowing up. Case quit the business in 1924 after producing just 132 steam engines in four sizes from 15 horsepower to 80 horsepower.

Huber Manufacturing Co.

Huber was among the most prolific builders of steam traction engines, and is one of the few manufacturers still in business today, right where they started in Marion, Ohio. The company made 11,568 engines, all apparently with the return-flue boiler design. Sales of steam traction engines stopped in the 1920s but the company still makes construction equipment for the same sort of market their steam rollers served 75 years ago.

"When I was 3 years old, in 1911, my dad built up a big brand new house on a piece of property he owned a few miles from the old place. Well, of course, we had to move the barn. At that time the tractor on our farm was a 35 horsepower double cylinder Buffalo Pitts—a good engine, and Murray Leonard was the engineer. I stood off to one side while they backed the tractor up to the barn—already loaded on wheels—and they hooked the chain to the drawbar. Well, they got the barn moved off to the new site . . . and nobody could find the kid! So they roared back across the prairie, hell-bent-for-election, looking for a grease-spot on the road where the kid had been. Finally, somebody had the sense to look in the barn, and there I was, asleep in the feed box."

—Ellis Nelson

1906
ADVANCE THRESHING MACHINERY
ADVANCE

CHAPTER THREE

ANATOMY OF A STEAM TRACTOR

A steam tractor is a sophisticated system with several major component systems: boiler, engine, drivetrain, and controls. It was, when it first appeared on the farm, the most complex and expensive machine farmers had ever seen—and a lot of farmers didn't like it. It set fire to things, blew up unexpectedly, and broke through bridges. But once farmers got used to the idea, and once they started learning (from their more adventurous neighbors) what steam power could actually do on the typical farm, they wanted to know more. So here is an abbreviated, de-hydrated version of the lessons those farmers learned when the steam tractor came to the American and Canadian farm.

Frame and Undercarriage

The earliest steam engines for farm use used wooden frames as a foundation for the heavy boiler assembly. Wood was a excellent material for the small engines of the time. Iron and then steel soon replaced wood, although cracks and breaks were common enough on steam tractor frames—and difficult to repair—during the whole period of steam tractor production.

The Woolf Compound patent was a standard fixture on Port Huron steam traction engines, a system that uses the same steam twice, once by a high-pressure piston, then again by a low-pressure piston.

OPPOSITE
The Corona Ranch, near Temecula, California, hosts an old-time farming "play day" in April, complete with the R&R Boys, Sonny Rowlands and Orman Rawlings, and their glittering LPG-fired Advance steamer.

Most engines, though, dispensed with the frame entirely and used the boiler as the mechanical foundation for the vehicle. Since this component was a carefully crafted structure fabricated from the best 3/8 inch sheet steel available, the design worked pretty well. It eliminated the weight of the frame, and that was important. But it also added some stresses and strains on the boiler which had to withstand both internal and external stresses.

The rear drive wheels normally attach to the sides of the boiler's firebox with large, bolt-on brackets with integral axle shafts. Since the loads on these brackets and shafts can exceed 10,000 pounds at a dead stop, and far more when the tractor is moving, these brackets had to be huge, heavy, well-crafted parts. Since no springing or suspension was provided for most tractors, a bump in the road could induce tremendous shock loads on the axle and its support. And a drop through a flimsy bridge might not destroy a sturdy tractor, but it could break off the wheels when the heavy, rigid machine suddenly impacted the ground below.

The front wheels normally attached to a bolster on the forward underside of the boiler. This attachment was very much like the pivot used on railroad cars

Drive wheels and their supporting brackets or axles must handle tremendous loads and stresses. As a result, a variety of designs and styles were available both from the original manufacturer and from "after-market" vendors. The 10ft drive wheels on the Best 110hp use 56 threaded rods, carefully adjusted and tensioned, to support the rim.

The front axle assembly on this Russell is pretty typical of steam traction engine design, a massive bolster and pivot supported by a heavy axle and simple steel wheels. Steering linkage on the Russell is the most basic type, a chain connected to a roller assembly.

and locomotives even today. While this bolster and the front wheel and axle assembly was pretty sturdy, the loads on it was far less than the rear wheels and the components were much smaller. No springing or suspension was provided on the typical steam traction engine front axle, and when you watched one of these machines rattle along a rutted dirt or gravel road you'd see why these components failed so frequently.

The boiler shop at Case looked about the same as any other builders, just a bit busier than most. This is how it looked in 1906, back when Joseph Fahey was the foreman. *J. I. Case*

Some steam traction engines, like the marvelous Avery undermounts, used a separate frame as the foundation for the tractor. Although this frame added weight to the machine it also offered some important design opportunities. On the Avery undermount, the engine and drivetrain mount directly to the frame, not on the side of the boiler as with conventional models. That meant that the power of the engine could be pretty much on line with the drivetrain, all the way back to the drive wheels, reducing some of the power loss associated with the geartrain needed with a top-mount engine.

Avery claimed, too, that the frame removed some of the stresses and strains on the boiler shell that inevitably occurred when it was used for support for the drive wheels, the engine, and the other major components. Avery was also proud of their use of the frame to get the engine down where it could be worked on conveniently—and there was plenty of work to do on an engine, so this was an important claim.

Cab And Controls

The cockpit of the typical steam engine can be just about as confusing, intimidating, and complicated as the "front office" of any fighter aircraft. But just as with an aircraft pilot, a tractor operator has many things to do and to monitor. The placement of the controls and their function had a lot to do with how the engineer used the machine to do the job—and what the engineer had to say about it, too. If the engineer didn't feel comfortable about the controls and the cab, a tractor wasn't likely to get a favorable recommendation. And since word-of-mouth is the most powerful form of advertising, steam traction engine builders invested

Port Huron; Here's what the engineer and fireman have to work with: (right to left) throttle, directional control, clutch, steam gauge (just off the peg) water ejector, whistle, and governor.

considerable attention in the science we now call "ergonomics."

As with a fighter aircraft, the typical steam traction engine requires a crew of two for efficient operation. The engineer's station is normally on the right, the fireman's on the left. The rear of the cab (if there is one) or the platform will ordinarily have two large bins or "bunkers" for fuel and water; these are like the fuel tanks of a conventional vehicle, but both fuel and water have to be fed into the engine's boiler manually—the fireman's full-time job. These bunkers on the Avery 40 horsepower undermount hold around 450 gallons of water and 500 pounds or more of coal, although these quantities vary tremendously with the size of the steamer.

Here are the controls found in a typical steam tractor and their functions:

Steam Pressure Gauge

A steam gauge was the primary instrument used by the engineer and the fireman. It was normally mounted smack-dab in the middle of the boiler, right at eye level. That was a good spot, because the engineer and fireman looked at it 1,000 times every working day. This gauge was, in a way, a kind of "fuel" gauge. The amount of steam pressure in the boiler was the factor that limited the work that could be done with the engine, not the amount of coal or water in the bunkers.

The fireman watched the steam pressure to anticipate fuel feed requirements. "The old-timers who taught me to fire a steamer believed you should keep the pressure at the same level, not let it go up and down very often," Glen Christofferson says. "Every time the boiler is pressurized, then de-pressurized, you cycle the stresses on the metal. The old-timers say you should keep the pressure up near the top of the gauge all the time."

Throttle

The throttle was usually a prominent lever with a locking latch mechanism

Best 110hp; the same standard controls and components are installed

right in front of the engineer's station. The throttle operated a simple valve feeding steam from the boiler to a line that ultimately terminated at the valve on the engine. This was the engineer's primary speed control, although "full throttle" provided different power levels, depending on the pressure of the steam available and the adjustment of the governor.

Try Cocks

At least two (and often four or more) small valves on the boiler allowed the engineer and fireman to physically check the actual level of water in the boiler by "trying" or testing. The upper try cock was placed above the normal level of the water in the boiler, the lower try cock below the lowest safe level, the crown sheet. Since the "sight gauge" could provide inaccurate information about the true water level in the boiler, the try cocks provided a means of verification: the upper cock should always vent only steam, the lower one only liquid water. Water out of the top meant the boiler is overfilled; steam out of the bottom meant you were about to be knocking at those 'Pearly Gates.'

Injector Control

The water level in the boiler was one of the critical concerns for the engineer and fireman. This level was adjusted manually, normally with the injector control typically found on the left side of the boiler for the engineer's convenience. This control allowed introduction of boiler feed water to replace that lost in the exhaust. The injector was one of two systems for replacing water in the boiler, the other being the feed pump described below.

Gear Shifter Lever

Most steam traction engines didn't have more than one gear, but for the ones that did, this lever permited selection and use of multiple gear ratios. A clutch control allowed the engineer to disengage the engine from the drivetrain just as with conventional vehicles. The

Gold Medals

Won by

Case Steam and Gas Engines

Winnipeg

International Contest

July, 1912

Construction. All riveting is done by hydraulic pressure. All holes tapped in the boilers are reinforced by an extra thickness of boiler plate riveted to the inside. The upper portions of rear head and of the front tube-sheet are supported or tied together by "through-stays" which extend lengthwise through the steam space and connect these heads. The "through-stays" are from 7/8 inch to 1 1/8 inches in diameter and from one to six in number, depending on the size of the boiler. The excess area of the rear head is supported by diagonal stays.

The Barrel of Case Boilers has no transverse seams. On the 60 horse-power size, and smaller, the longitudinal seam is double riveted; on the 75 and 80 horse-power it is triple riveted; and on the 110 horse-power size, a double butt strap is employed and triple riveted on each side. The barrel is reinforced for the front bolster and around the hole under dome.

The Dome is reinforced about all openings, and has sufficient capacity to always supply dry steam to the cylinder. A lock pop safety valve is attached to dome.

The Fire-Box is large enough to insure proper combustion of fuel without forcing the fire, and the long tubes are so proportioned that they extract nearly all the heat from the fire before the burnt gases escape into the smoke-stack. The open bottom (not water bottom) is used. The fire-box sheets are flanged to meet the outer sheets, thus omitting entirely the objectionable mud-ring, and making one joint instead of two.

The Crown-Sheet is stayed to outer sheet in the manner common to locomotives. Stay bolts of special double refined iron, 7/8-inch diameter, are used on all engines except the 110 horse-power, in which they are one inch diameter, spaced close enough to pass the most rigid examination. The heating surface, grate surface and size of cylinder are proportioned to give best results. The crown-sheet is fitted with a fusible plug.

The Side Sheets are extended down and form the sides of the ash-pan. To the sides of the boiler are riveted extra heavy wing sheets, carrying the engine, gearing, etc. The weight of the boiler is carried on springs resting in pressed-steel brackets, held to the boiler

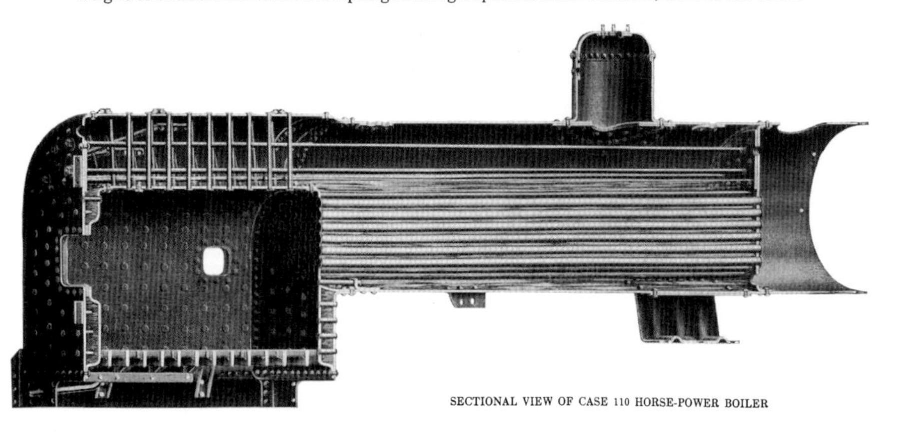

SECTIONAL VIEW OF CASE 110 HORSE-POWER BOILER

Gold Medals. Case 1913 catalogue description. *J. I. Case*

Westinghouse traction engine used a simple slider to change gears without benefit of a clutch.

Reverse Lever

The reverse control was another one of the engineer's controls typically located on the right front quadrant of the cab or tractor platform. Unlike the reverse gear in cars, trucks, and tractors, a steam engine reversed rotation in order to back up. The engineer's reverse control changed the timing of the steam valve. All steam tractors had a cutoff valve that allowed you to rumble down the road at top speed and shift from forward to reverse without even bothering to stop. The tractor would gradually come to a stop, then start moving backwards. Most steamers would go just as fast (or slow, depending on how you

Avery steam traction engines normally used this "undermount" placement for the cylinder and associated engine components. Down here, close to the ground, the engine is much easier to work on. The drive train is simplified, too. But the mechanical components are also much closer to the dirt of the field, a real potential problem during plowing.

look at it) in reverse as forward in each available gear ratio.

Water Gauge

The water gauge was one of five safety provisions (along with the fusible plug, upper and lower try cocks, steam pressure gauge, and safety valve) normally found on steam traction engines. Along with the steam gauge, it was consulted by both the fireman and the engineer 100s of times in a working day. It was really a simple instrument, no more than a glass tube mounted vertically. Plumbed fittings allowed the water level to be displayed in the tube—as long as the water was within the safe range. The fireman and engineer both monitored the water level in this gauge (also known as a "sight glass"), adding water as required.

ABOVE AND BELOW
The linear motion of the piston is converted to rotary action by a "pitman arm" or crankshaft (depending on which manufacturer's literature you're reading. This engine is turning very close to 250rpm.

Water Pump Control

There were two ways to introduce water to the boiler, both controlled from the cab or platform. The injector pumped water at a constant rate; the manual feed pump was variable. Of the two, the manual feed pump provided a bit more control. On Russell tractors like the model owned by Glen Christoffersen, the water pump worked continuously and either recirculated the water to the tank or sent some to the boiler, depending on the water pump control valve position. "The feed water pump control on my Russell provided better control than the injector," Christoffersen reported.

Firebox Grate Lever

By 1900 or so, convenient controls allowed the fireman to adjust the grate position and dump the accumulated ash and clinkers with a lever on the rear of the firebox. Not all steam traction engines had this feature, but the crews of the ones that did certainly appreciated it.

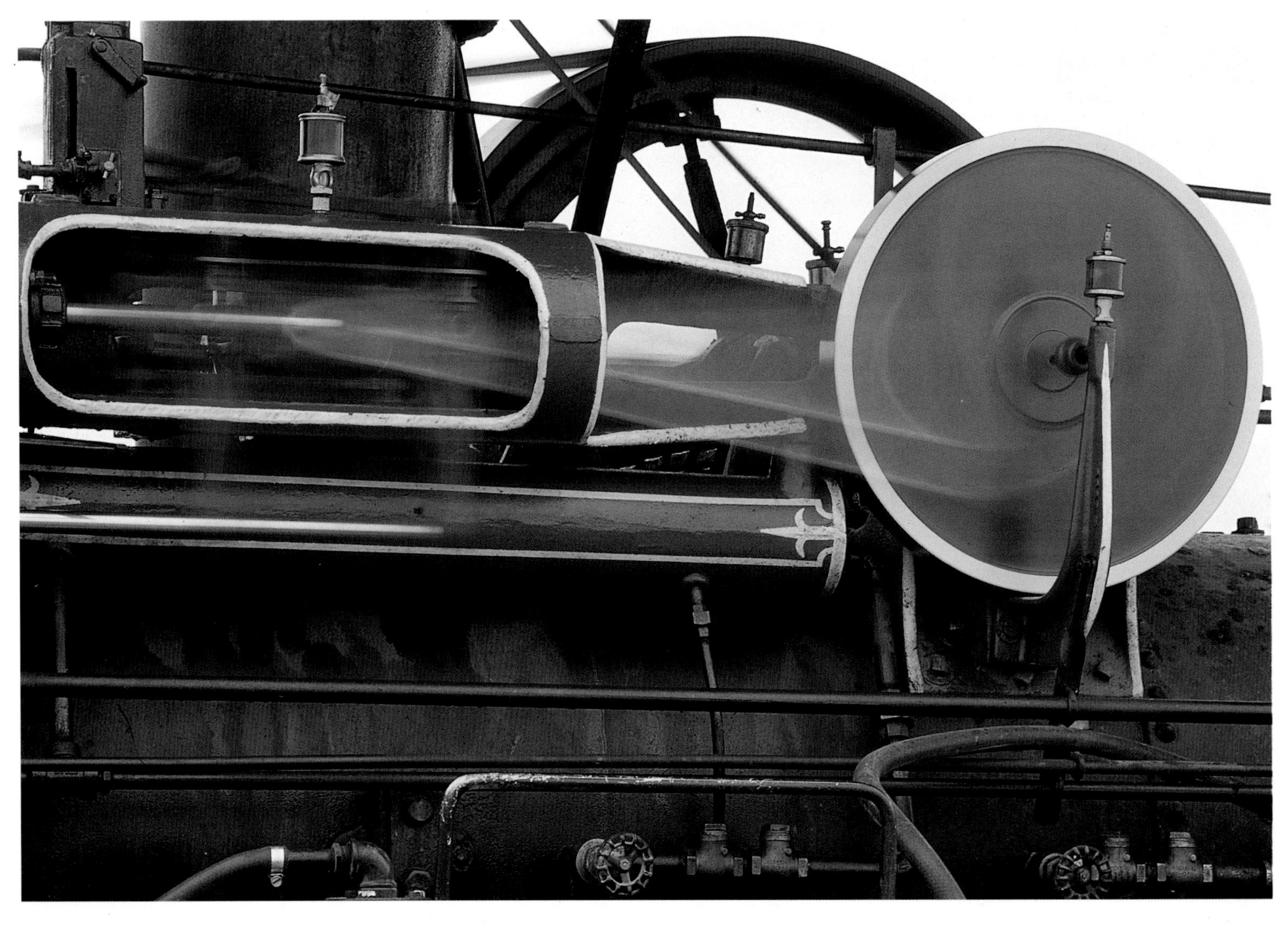

Ash Dump Foot Control

Another innovation for the ease of the crew was the provision for dumping the accumulated ashes—and there were plenty of them, particularly with wood-fired engines. These ashes needed to be cleaned out, normally several times a day. While most steamers required the fireman to clamber off the tractor and clean out the ash pan with a shovel and rake, the Avery and some others dumped the material with a simple press of the pedal.

Fuel Bunker and Water Tank

Steam traction engines consumed huge quantities of fuel—coal, wood, straw, or oil. The first three of these were bulky, but cheap, and were the standard fodder for all of the engines shown in this book (although one has been converted to liquid petroleum gas (LPG)). A hard-working engine might have consumed up to 3,000 pounds of coal in a single day, and most would have gone through 100s of gallons of water. Both water and fuel needed frequent replenishment so ample storage was provided on virtually all steam traction engines, even those that routinely pulled a tender with additional supplies.

The coal bunker was normally on the right side of the platform, at the rear, and typically held between 300 and 500 pounds of coal. Wood-burners sometimes configured this bunker differently; the Best 110 horsepower owned by the Oakland Museum had stowage for fuel overhead, a long reach for the fireman.

Water capacity was typically between 200 and 450 gallons, normally in a tank at

There are many of these automatic oilers on this Avery and most other steam traction engines. Each must be checked and filled several times every working day.

BELOW

Three of the dozens of oilers and "hard oil" cups on the Avery 110hp. In addition to gravity and simple pressure feed of lube oil and grease, many steam traction engines used high pressure pumps, of either mechanical or hydrostatic design, to force oil into critical bearing surfaces.

Gold Medals, page 2. Case 1913 catalogue description. *J. I. Case*

COMPRESSED SECTION OF TUBE SHOWING DUCTILITY

SECTION OF TUBE FLATTENED

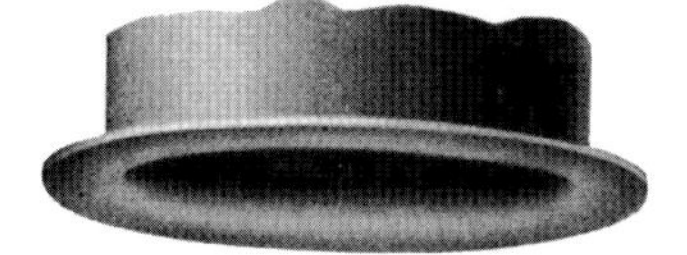

END OF TUBE FLANGED

by rivets and by bolts below the water-line. By our method of mounting the engine and boiler no important part is held to the boiler by bolts tapped into the steam or water space.

The Smoke-Box is from 19 to 28 inches long, depending on size of boiler.

Boiler Tubes. The boiler tubes are made from cold-drawn, seamless steel, soft and ductile in quality, which makes them easily expanded and beaded. They are 2 inches in diameter in all sizes excepting 18 and 30 horse-power engines, in which they are 1¾ inches. They are arranged in vertical rows, which insures free circulation of water and permits sediment to settle on the bottom where it can be washed out through the hand-holes. They are easy of access for cleaning. The tube holes are in perfect alignment in the tube-sheets and are carefully reamed, so that the tubes fit accurately and are easily kept tight.

Hand-Holes. There are six washout holes for cleaning the boiler, located as follows: One at each corner of water leg, one below tubes in front tube head and one in rear boiler head above crown-sheet. Pressed-steel hand-hole plates are used, thus eliminating the old-style cast-iron ones. In addition to the hand-holes there are a number of washout plugs in the sides and front of fire-box and front tube-sheet, and in top of barrel, back of the dome.

Straw-Burning Boiler. Substitute a fire-brick arch and short grates for the coal and wood grates, attach the straw chute with flaring outer end to the fire-door and you have a straw burner. The straw chute enables you to feed the straw easily into the furnace—the trap door inside of same excludes the air when the chute is not filled with straw. After the boiler has been fired for a short time the brick arch becomes white hot, which keeps the fire-box at an even temperature and insures perfect combustion and uniform evaporation. Because of the dead plates the draft enters the fire-box at the front end and is drawn against the straw as it is forced through the door, thereby securing rapid ignition. The CASE Straw-Burning Boiler may be changed to burn wood or coal by removing the fire brick and the straw chute and substituting the regular fire-door and coal grates. If so desired a part of the fire brick may be left in place when coal is used as fuel.

Jacketing. The jacket consists of wood lagging covered with hammered, polished steel, with brass bands at the joints. A jacketed engine is more easily kept free from dirt and oil, and the slight additional cost is more than saved by a prevention of heat waste, especially in northern climates. An extra charge is made for the jacket on all engines, except the 110 horse-power and the road roller.

FIRE-BOX OF STRAW-BURNING BOILER

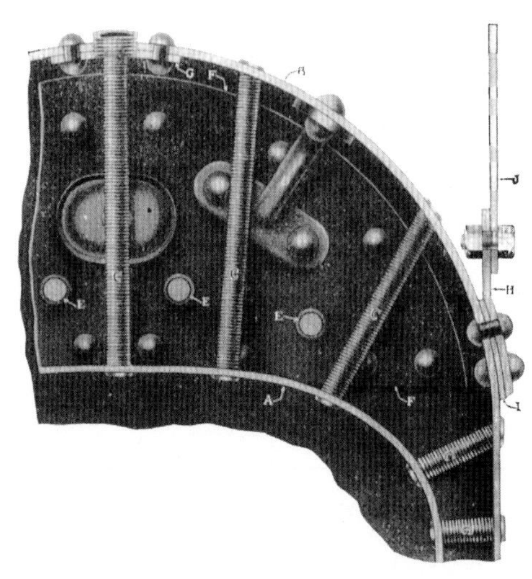

PORTION OF WAGON TOP AND CROWN-SHEET

Steel-shell Feed-Water Heater. Case 1913 catalogue description. *J. I. Case*

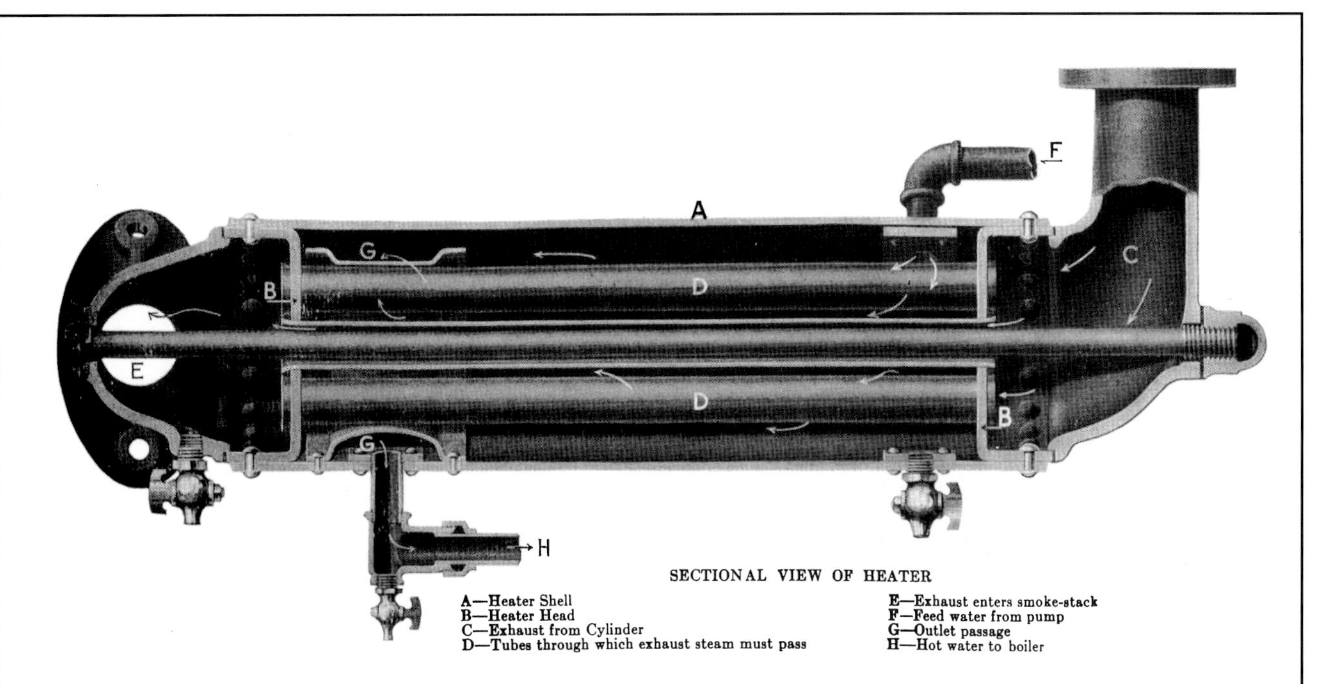

SECTIONAL VIEW OF HEATER

A—Heater Shell
B—Heater Head
C—Exhaust from Cylinder
D—Tubes through which exhaust steam must pass
E—Exhaust enters smoke-stack
F—Feed water from pump
G—Outlet passage
H—Hot water to boiler

Steel-shell Feed-Water Heater

IN a large number of the testimonials from users of CASE Engines you will find the statement: "She is an easy steamer." There are many reasons that contribute to this desirable feature of the CASE Engine that may be found in the general design of the boiler and fire-box, but the Water Heater plays a very important part. The exhaust steam is utilized to heat the water before it goes into the boiler. This heater was designed and is built by us. The feed-water surrounds the tubes, and the exhaust steam passes through them. The hot water is discharged to the boiler on the under side of heater, but is taken from the top, the water passing behind an annular flanged plate that leads to the outlet. The interior is readily accessible by removing the exhaust inlet and outlet elbows.

The Case Geared Pump is now used on the 40, 50, 80 and 110 horse-power engines. It is driven by a gear on the crank-shaft, meshing with a larger gear that carries the crank-pin which operates the pump. This pump has sufficient capacity to supply 50 per cent. more water than is needed by the boiler in extreme conditions. The feed water on its way to the boiler passes through the steel-shell heater. By means of a by-pass valve, the water can be allowed to return to the tank when not needed for supplying the boiler. The pump is, therefore, always ready for instant use.

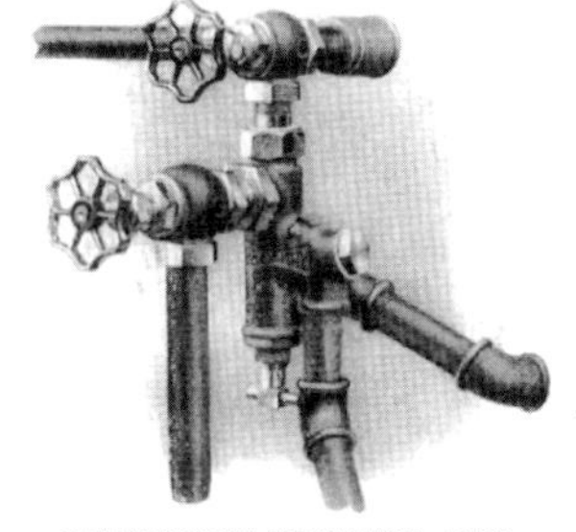

PENBERTHY INJECTOR AND FITTINGS

Injectors. All CASE Engines are equipped with the well-known and highly efficient Penberthy Injector.

GEARED PUMP

Steering gear on the huge Avery 110hp. This worm and follower system is far more sophisticated than the chains used on most steam traction engines.

The cylinder on this Russell is not much more than a large, simple, very solid tube of cast steel. An insulated blanket, protected by this sheet metal cover, helps retain heat and prevents condensation of the 350 degree steam.

the left rear of the platform. The Best 110 horsepower demonstrated an alternative here, too, with a horizontal tank at the front of the machine, right where a "locomotive-style" boiler normally appeared.

Boiler

The foundation for the whole business was boiling hot water, all contained in a big, heavy "can" made from thick (normally about 3/8 inch) sheet steel. A 60 horsepower engine would need 250 gallons of water in the boiler all the time, a large volume and a very heavy mass. One cubic inch of water, when evaporated, would expand to form about one cubic foot of steam. That expansion could be controlled and used to move a piston. The piston's linear motion could be easily converted to rotary motion, and that motion could do work. A 1917 Case 60 horsepower engine's boiler would carry a bit over 1,500 pounds of water.

The power of a steam engine was entirely dependent on its boiler system—the grate, the flues, the water capacity, and the space above the surface of the water where the steam accumulated. A typical tractor boiler would have between 36 and 76 tubes, each about 2 inches in diameter. The common tractor used horizontal flues, but many earlier machines used vertical boilers; both worked, but the locomotive style was better suited to the larger engines, had a lower center of gravity, and better clearance under obstacles.

The long horizontal portion of the boiler was the *barrel*, and it contained the *tubes*, or *flues*. On top of the barrel was the *steam dome*, a kind of reservoir for the accumulated steam. The flues were essentially long, (usually) 2-inch steel tubes supported by the *front flue sheet* and the *waist sheet* at the rear. A typical 60 horsepower Case simple single cylinder engine used 40 of these 2-inch tubes, each 90.5 inches long. For a variety of reasons, the flues weren't welded in, but retained between the front flue sheet and the waist sheet by

Flywheel and intermediate gearing. The clutch shoes are visible through the spinning web of the huge flywheel. Sycamore, oak, and other woods each have their proponents—but you've got to replace those shoes pretty often if your steamer does much work at all.

mechanical fit—their ends were expanded and rounded over with flaring tools. A copper gasket helped form a water-tight seal. This allowed individual tubes to be easily removed and replaced by first cutting in the middle (with an internal tube cutter), then each half could be pulled out from the firebox or the smokebox ends.

The rear portion of the boiler was typically a boxy shape—the *firebox*. The heart of the boiler, and by extension the whole tractor, was the firebox. This space contained and directed the heat of the burning wood, straw or coal. A typical engine might have had a firebox about 42 inches deep, 30 inches wide, and 27 inches high. That's a tight squeeze, but the engineer or mechanic would have to work inside from time to time, replacing flues and grates. The *grates* supported the burning fuel, admited air and allowed the ashes to drop down to the *ash pan*. Access to the firebox was through the firebox door, at the back of the boiler. *Inspection doors* on each side of the firebox allowed the engineer to peer inside without admitting a lot of cold air to disrupt the fire. The 1917 Case 60 horsepower engine mentioned earlier had a grate

At 40psi the big Best vertical boiler is just starting to simmer. It has been warming for an hour and it will be another hour or so before the engineer will have full pressure up.

When firing with wood, the firebox is kept pretty much jam-packed full all the time. The fireman is busy all the time, too, keeping it stoked and the ashes under control. A small "weeping" leak from one staybolt (above the firebox door) will seal up as the metal expands and the pressure grows.

area of just under eight square feet; it heated an area of 216.9 square feet.

The heat from the burning fuel produced firebox temperatures of over 1000 degrees Fahrenheit—enough to heat the steel red hot if it were not for the surrounding water. In fact, the water surrounded the whole firebox, including most of the back surface, except for the firebox door.

Boiler Styles

Two basic types of boilers were found on farm steam traction engines: the "locomotive" type and the "vertical" type.

"Locomotive-style"

Farm steam tractors typically used a railroad locomotive-style with "fire in tube" boiler design—a layout with the firebox at the rear, horizontal flues, and an exhaust in the front. A variation on the concept routed the heat through one set of flues to the front, then back to an exhaust at the rear through a second set of flues; this type of boiler attempted to get a little more heat transfer from the fuel's combustion gases and was called a "return flue" boiler design. The return-flue was never as popular as the direct flue boiler but you will still see plenty of them at the shows, easily recognizable by their smokestacks in the middle or rear of the boiler's upper surface.

"Vertical-style"

Not all steam traction engines used the "locomotive" style design. Some, including those from Best, Westinghouse, and others, used a vertical boiler system with its own vices and virtues. Vertical boilers were normally more compact and often permited better visibility for the engineer. The water in the boiler didn't slosh back and forth on grades as much as the horizontal boiler. Westinghouse and other builders of vertical types claimed advantages for this layout that most farm engineers never quite discovered. Westinghouse, for example, claimed its boilers would make 100psi steam from a cold boiler in just 20 to 30 minutes, that fuel consumed was a third to a half that needed by "locomotive" designs of the competition, and that actual power produced by its 18 horse-

Drive shaft and intermediate gearing on the Best 110hp.

Early morning at Rollag, Minnesota, on Labor Day weekend finds a huge congregation of old steam traction engines and the men that love them ready to work once more.

power engine was actually 42.5 horsepower. Maybe so, but you'll see very few of these Westinghouse tractors in shows or museums—farmers apparently weren't convinced and bought Case, Russell, Advance, and Minneapolis horizontals that outnumber the verticals about 50 to one on the roster of survivors.

Boiler Components

The roof of the firebox was a large, heavily reinforced section of steel plate called the *crown sheet*. This was the most vulnerable, critical component in the steam engine; when a boiler exploded the crown sheet was usually the source of problem. A "fail-safe" device installed in the crown sheet, called a *fusible plug*, protected the engine by melting when the water level fell below this portion of the firebox. When that happened, a rush of steam entered the firebox, putting the fire out and releasing the pressure on the boiler.

The heat from the fire in the firebox was absorbed by the sides of the firebox, by the crown sheet overhead, and out through the flues to the *smokebox* and then up the *stack*. The design of the system was intended to convert as much of the heat from the fuel to steam as possible, but a tremendous amount was wasted.

Plumbing

The very first steam engines used the condensation of steam as part of the power-generating process, but condensation and cooling on steam traction engines was a problem rather than an advantage. Any cooling of the 300+ degree steam between the boiler's steam dome and the engine's exhaust would reduce the amount of work the engine could accomplish. This was really important on frigid days on the northern Canadian prairies when the wind blasted down from the Arctic, chilling the men trying to

Engine Mounting and Gearing

THE sectional cut herewith illustrates how the boiler of the CASE Engine is suspended on springs and carried in front of the rear axle and counter-shaft. This is a notable feature of the CASE Engine. In order that you may better understand this feature of construction: Suppose the axle of your wagon was run through the box, would not the wagon feel every light as well as heavy jolt while being driven over the road, and would not the life of the wagon be shortened? The traction wheels and gears of CASE Engines are mounted independent of the boiler and fire-box by means of radius links, which connect with the counter-shaft. Steel spring-pot brackets, resting on spiral springs, support the weight of the boiler. These springs are suspended from the lower cannon bearing, which encases the axle, on eye-bolts which allow the springs free play. The distance between the counter-shaft and rear axle is maintained by distance links, provided with turnbuckles, which allow the gears to be kept in perfect mesh. These links permit an up and down movement of the boiler without in any way disturbing the mesh of the gears or the equilibrium of the boiler, and without subjecting either to shocks or strains. Side play of the boiler or counter-shaft is prevented by means of a steel cross link held to the rear of the boiler by a stud bearing, which is fitted over a trunnion of the upper cannon bearing. On the lower cannon bearing is a heavy lug, which is kept in position by parallel guides riveted to the boiler below the fire-door. This arrangement, with the cross link above prevents all side or end play. Our system of spring mounting is not only theoretically correct, but for many years has proven thoroughly efficient on engines doing all kinds of road or contract work and plowing. In addition to the spring mounting of the boiler and spring differential, the platform is mounted on springs, the draw-bar has a good strong one, and each of the guide chains has one to give elasticity and to keep the chain taut.

VIEW OF RADIUS LINKS, DISTANCE LINK AND SPRINGS SUPPORTING BOILER, SPRINGS UNDER PLATFORM AND IN DRAW-BAR

VIEW OF BEARINGS OF COUNTER-SHAFT AND REAR AXLE, LONG HUBS OF TRACTION WHEELS, DISTANCE LINKS AND SPRINGS SUPPORTING BOILER

Drive shaft and intermediate drive train components, Avery 110hp.

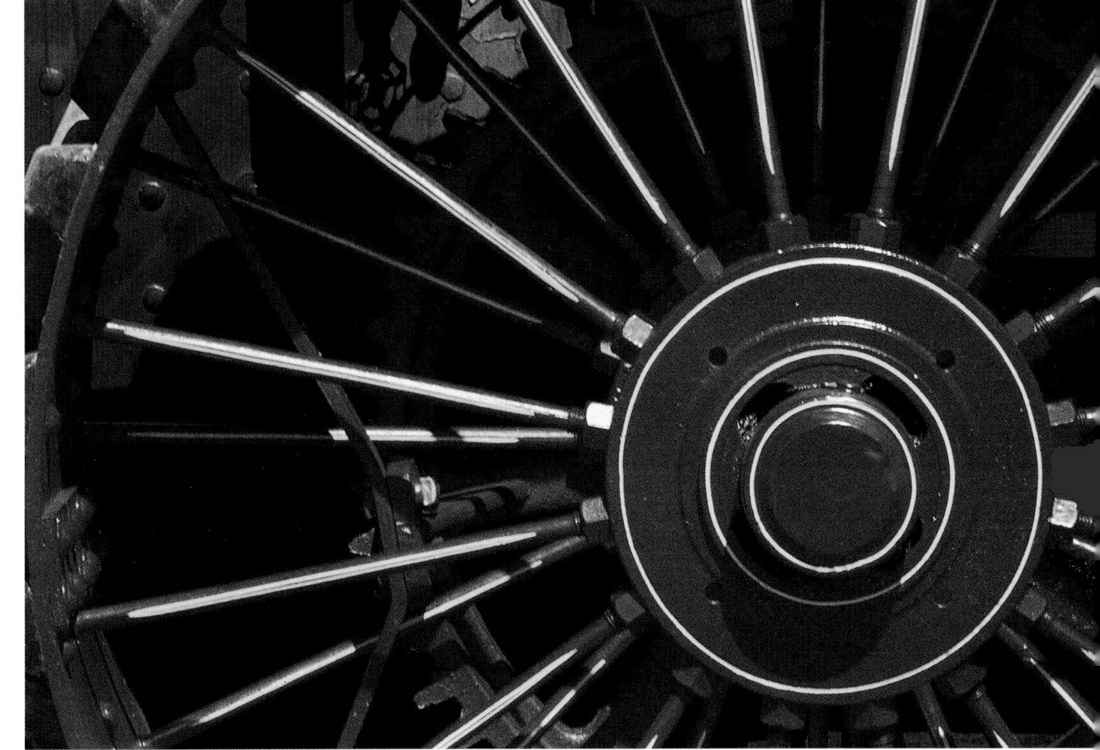

The drive wheel on this Russell tractor is about 6ft in diameter. Behind the spokes you can see a heavy cast iron gear with its attaching brackets. Breakage of these big gears was fairly common, despite their massive construction.

LEFT
Engine Mounting and Gearing. Case 1913 catalogue description. *J. I. Case*

thresh a late grain harvest in October or even November. So to keep as much of the heat within the system as possible, builders often put jackets on the boilers and on the steam lines to retain the heat. Some boiler designs used a double shell to provide some of that insulation.

But a crucial part in the machine under these circumstances was the steam line itself. Normally this was a 2

That's the steam dome on the right, the source of the dry steam used for the engine—and for all those whistles, too, that have been installed on this nice Advance Rumley engine. The device with the spheres is the governor, set to keep the engine to no more than 250rpm.

Avery often supplied steam traction engines with what was commonly called a "Canadian" boiler. That design featured extra strength (and extra rivets) for higher pressures than commonly used in the US. That's the steam dome on top and the feed water heater (the horizontal cylinder with the Avery legend on the side) along the left portion of the boiler.

inch or 2.5 inch pipe, leading from the steam dome valve to the throttle. The large diameter of this steam line reduced friction, and the steam line was designed to be as short and straight as possible. Normally, too, it would be routed close to the boiler shell to retain heat and may have had a jacket of insulation of its own.

Engine

The pressurized steam from the boiler was piped from the steam dome, by way of a throttle valve and governor, to the engine proper and a lube oil injector.

A steam engine was really a very simple device, particularly in its most basic form. A throttle system, including a governor, allowed the engineer to control the speed of the engine. Again, using the 60 horsepower Case as an example, the engine was designed to operate best at 140psi working pressure.

That 140psi made the Case a high-pressure system; other engines worked at much lower pressures, sometimes down around 20psi. The problem with low pressure steam was that it wasn't a very good use of the fuel. It only took a little more fuel to raise the pressure in a boiler from 80psi, for example, to 120psi. A high-pressure system used fuel better, saving money and effort.

This was particularly true with the "compound" engine that used high-pressure steam twice before allowing it to vent to the atmosphere. If it took 100 pounds of coal to bring the water in a high-pressure boiler from 60 degrees Fahrenheit to steam at 5psi, it would take another 103 pounds (a total of 203 pounds) to bring it to 80psi or 104.5 pounds (a total of 204.5 pounds) to bring it all the way to 160psi! Only a pound and a half more coal will provide a huge increase in steam pressure, and that pressure could do a great deal more work—a 15 to 30 percent advantage, if the engine was run at a constant full load.

TRANSMISSION GEARING

Power is Applied Directly to the Tires of Both Traction Wheels. The CASE Engine is noted for its great pulling power and its ability as a hill-climber. Why? It is not all due to the power of the engine. It is due to the way in which the power is applied to the driving wheels. In order to make this point clear, we will suppose that we take two driving wheels attached to the counter-shaft and suspend them on bearings a few inches above the ground. Then we will ask you to turn the wheels by a grip on the counter-shaft. This would be almost impossible to do. Next we would ask you to turn them by taking hold of the spokes near the hub. It would be easier to turn them from this point, but the power you would have to apply would be still great. Again we would ask you to take hold of the outside rim. From this point you could spin them around with ease. Now you understand what we are driving at. In CASE construction the power is applied direct from the extra heavy rim of the bull gears to the tires of both traction wheels. How this is done is illustrated by the cut at the top of this page. You will notice there are eight bars leading from the rim of the bull gears direct to the tire of the traction wheel. This brings the pull to the outside of the wheel where the leverage is greatest.

The Traction Gears are made from CASE special ferro-steel, cast in our special gear foundry and prepared from specifications made in our laboratories. We have used this material for many years on all our engines, and the records show there has been less wear and breakage than on any other gear material used for this purpose. Two pinions and two gears are used to transmit the power to the traction wheels.

DIFFERENTIAL GEAR WITH COUNTER-SHAFT PINION REMOVED

Spring Differential Gear. CASE Engines are equipped with a differential gear in which a series of coil springs receives the impact of sudden starting and gradually transmits power to the driving gears. The differential itself is very essential, in that it facilitates turning by allowing the traction wheels to accommodate themselves to the different distances they travel on curved roads or in turning. It may be emphasized at this point that both rear wheels of CASE Engines are drivers at all times, going either forward or backward on a straight or curved road.

REAR TOP OF BOILER SHOWING WING SHEET BRACES AND GUIDE FOR CANNON BEARING

The Steering Gear is fitted to the front end of fire-box with strong brackets which hold the chain roller in place. The chain roller is cast in the form of a right-hand auger, making a guide channel for the chain to prevent it from crowding or overlapping. The steel chains are supplied with springs, and by means of our method of attaching to the front axle have the same leverage in turning the front axle whether the wheels be straight or cramped to the utmost. With this arrangement the chains are kept to the same tension, and guiding of the engine is done with comparative ease.

LEFT
Engine Mounting and Gearing, page 2. Case 1913 catalogue description. *J. I. Case*

Water Injector (or "Ejector")

Terry Kubicek is a Nebraska steam fan with a collection of steam traction engines and lots of experience as engineer and fireman on Keck-Gonnerman, Case, Nichols & Sheperd, and other types. Terry explains one of the most sophisticated (and troublesome) components of late-model steam traction engines, the water "ejector"—often called the "injector."

"An ejector is used for lifting water or filling tanks and can be called a siphon or a jet pump. It operates by piping steam into a chamber where it intercepts a cold water pipe and by a the force of the steam siphons the cold water with it into an orifice and thence to a discharge pipe that fills the water tank on the steam engine. An ejector is not used to fill the boiler on the steam engine, but is used to fill the water tank.

"Water is then moved from the water tank by a geared pump operating as the engine is engaged, or by use of the injector. The injector operates independently of the engine and found at the junction of the steam pipe extending from the boiler steam dome and the cold water pipe leading from the water storage tank. The injector operates with a set of valves: a steam valve and a cold water valve.

"There is a check valve between the steam dome and the injector and

It looks like a flywheel but that part is on the far side of the boiler; this is the "crank disk" and is just as big and heavy as a flywheel on a smaller engine. But its strength and weight only help transmit the power from the engine's pitman arm to the crankshaft.

The cylinder and engine assembly on this Advance Rumley uses massive castings both for strength and for thermal efficiency. A steam engine runs best when it is hot, just as hot as the steam coming from the boiler. All that cast steel holds heat and minimizes condensation.

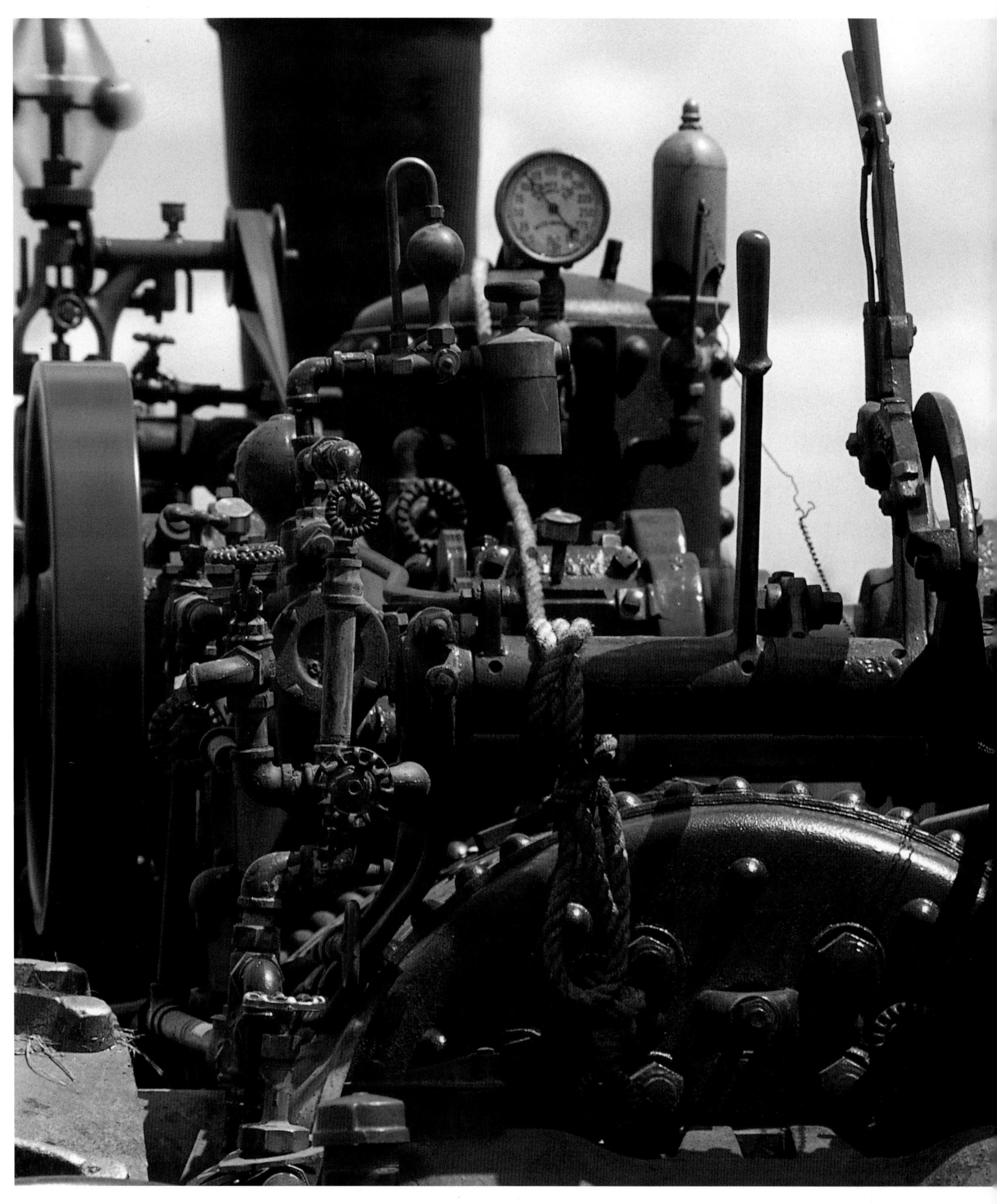

Pitman arm, crank disk, and grease cup on a Port Huron engine.

Cockpit layout for the Port Huron.

often there is a check valve on the feed water pipe between the injector and the middle part of the boiler. As the cold water valve is slightly opened, the steam valve is opened and steam travels through a tapering nozzle to a small chamber intersected by the water supply pipe. As the steam enters the chamber a partial vacuum is created that causes the cold water to siphon into the chamber.

"As the steam picks up the cold water, the mix increases in velocity and mass as it shoots through another nozzle and finds the discharge pipe through a check valve and into the boiler. As the volume of the steam and cold water mix can be regulated by the engineer, the overflow valve may open and discharge hot steam and water. If the overflow valve drizzles and keeps the injector hot, the injector may not "take up" the cold water. Playing with the degree of opening of the steam valve and the cold water valve may remedy the situation. However, if it does not, then the injector must be cooled off and kept cool as the engineer manipulates the steam volume and cold water volume. If the injector refuses to function, then the orifices, the internal jets, or the valves themselves may be worn, or blocked by either lime or debris."

Engine Mechanism

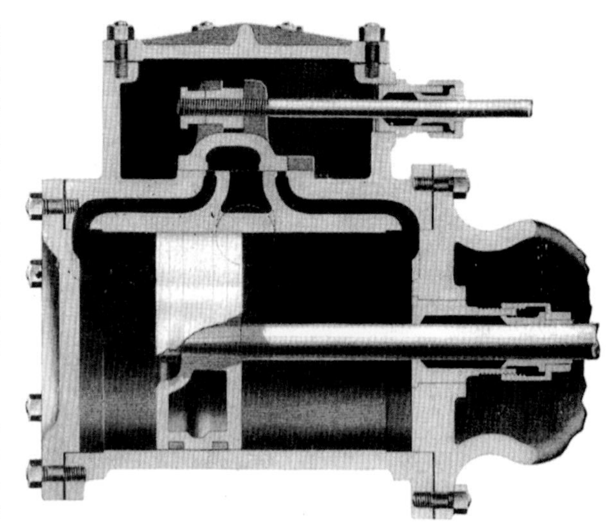

SECTIONAL VIEW OF SIMPLE CYLINDER

THE *Case Engine* is a single side crank of the simplest type. The frame is of the girder pattern, and the cylinder end is faced and the guides bored at one setting, so that they are in perfect alignment with the cylinder. Before removing the frame from the boring machine the babbitted main bearing is accurately bored at right angles to the central line of guides. Every part is easy of access for oiling or adjustment. The large disc and heavy fly-wheel give perfect balance to the engine and permit of its being run as slowly as desired.

The Simple Cylinder and Steam Chest are cast in one piece of special close-grained iron, which insures a smooth, durable and easily lubricated surface. To guard against boiler expansion all the cylinders of our engines overhang the frame, and are not bolted to the boiler or to a heater. The steam ports are of ample area to prevent "wire drawing," with all unnecessary clearance or waste space avoided.

The Slide Valve is the plain D style locomotive type. The valve and valve-seat on every engine are carefully machined to a true surface and then scraped by hand to insure a perfect steam-tight fit.

The Piston is a single hollow casting of sufficient width to give ample bearing and wearing surface. The piston rings are of improved form and self-adjusting. The piston rod is made of a selected grade of steel. The hole in the piston for the piston rod is bored to a standard taper and the piston head is forced on with a pressure of about twelve tons. As additional security a jam nut is put on and the end of the rod is then riveted over.

The Cross-Head is fitted with shoes, accurately turned to the same radius as the bore of the guides. The shoes may be easily adjusted by means of two screws at each end, so that the wear may be taken up and the piston rod kept in proper alignment.

The Connecting Rod is of the latest approved design of I-beam section, very strong and rigid, although light in weight. It is forged from a single piece of steel without welds. Both ends are of the box form, no straps, gibs or keys being used. It is made unusually long, about three and one-fourth times length of stroke, thereby lessening the angular thrust and reducing the friction between the cross-head shoes and guides. The connecting-rod boxes are made of phosphor bronze, the best anti-friction material obtainable.

CROSS-HEAD WITH ADJUSTABLE SHOES

The Crank Disc is of large size and properly proportioned to counterbalance the reciprocating parts. It is forced on the shaft by hydraulic pressure of at least fifteen tons, and afterwards a carefully fitted key is driven. The crank-pin is also pressed into the disc by heavy hydraulic pressure and afterwards riveted.

Hub assembly, Best 110hp. This wheel is about 10 feet tall and supports a static load of over 5,000 pounds. Two grease cups provide good lubrication and get reloaded daily.

Boiler fabrication was once an important art and a highly respected skill—and with good reason. Four rows of carefully placed and installed rivets, with a generous overlap of 3/8 inch steel rated at 60,000 pounds strength, and reinforced by many long staybolts will safely contain 200 pounds or more steam pressure.

A heavy tow chain was an essential accessory for steam traction engines, useful for both towing and for being towed.

This is the plain-vanilla steering system found on most brands of steam traction engines. The slack in the chain makes for a lot of slop, but you weren't going to try taking corners at high speed anyway.

LEFT
Engine Mechanism. Case 1913 catalogue description. *J. I. Case*

"D" Valve

The "D" or "cutoff" valve, also known as the *forward/reverse control*, regulated the timing of the introduction of steam pressure to the cylinder, controlling both the efficiency and the direction of rotation of the engine. The position of this control, and the valve to which it was attached, determined the power the engine would need to provide to go forward or backward; it also drove the belt off the pulley either *away from the cylinder* ("running over the top" in steamer parlance) or *toward the cylinder* (referred to as "running under").

The cutoff valve did more than that, though; you could "lean out" the operation of the engine in a way somewhat similar to the way a gasoline engine was adjusted to run as lean as possible, consistent with power output. The cutoff valve control regulated the timing and duration of the valve cycle; the leanest,

THE KID FROM KANSAS

By Ellis Nelson

"A Port Huron engine was shipped into our part of the country, sometime around World War I, a 25 horsepower with a single cylinder and a piston-type valve. This type of valve was a great improvement on steam locomotives around the turn of the century, and some companies adapted the idea to use on steam traction engines. The Port Huron company made one of their own, and this valve was on the 25 horsepower machine we were using there at Antler, North Dakota. It worked well enough on the railroad and when a skilled mechanic adjusted them, but those 'farmer-mechanics' we had up there had trouble with them.

"This engine finally needed to have its bearings repoured and a local engineer, Bill Balance, was hired to do the job. Bill was a good engineer normally, but he got drunk about the time he set out to re-babbit the bearings and got the lower bearing oversize! That made the crankshaft too high. And that made it just about impossible to adjust that piston valve to save your life!

"Well, about the summer of 1926 something went wrong with our old Avery and Pa had to have an engine, right away! He made a deal for that Port Huron, then sitting way outside town. He and some guys he'd hired to help with the thrashing went out there and steamed that Port Huron up, and I went along—I hauled the water for it.

"We brought that thing home and the first thing they tried to do was to set that valve to make it run right, but it couldn't be done without shimming the pedestal under the cylinder, which they tried. That got a pretty good result. We finished the thrashing in good shape with that engine.

"Then, the next year a kid in a Model T full of tools showed up—a young 'gaffer,' just 22 or so—drove in the yard one evening. He was from Kansas and was looking for a job as engineer on a thrashing crew. He *claimed* to be an engineer, but Pa wasn't too impressed with the package. But he was hungry and needed a place to sleep. Pa told him "Throw your stuff in the bunk car, and the wife will get you something to eat; you can be engineer, if you want."

"That kid took on the Port Huron, and he was *serious*! He reran the bearings and got the shaft right where it was supposed to be, right in line. He did everything just right. Within a couple or three weeks he had that steamer in fine shape. All he was getting out of the deal right then was a place to sleep and his meals—but my mom was the best cook in the county, so he made out okay on that deal. And my folks kinda got to like that kid."

"A neighbor down the road, Ole Henley, had a brand new, all-steel Nichols & Shepherd full-size threshing machine. For the use of the thresher for the season, my dad thrashed Ole's crop. So they went to work thrashing, that summer of 1927, and my dad said he never worked on a rig that ran as well as that one did! He said, 'That gol-durn kid from Kansas was the best steam engineer I ever had anything to do with! And that Port Huron, with the piston valve, must have had eyeballs because it could see a bunch of bundles coming, the way it took the load!' He said it was the smoothest running, best operating machine he'd ever had anything to do with."

Steam gauge.

The crosshead provides the same function as an automobile engine's piston pin, the part of the engine where linear motion is converted to angular or rotational motion.

Many components of every steamer were made of cast iron or steel, a fabrication process developed into a highly skilled craft by men like these 100 years ago. This is part of the foundry at the Case factory, circa 1910. These men might be pleased to know that hundreds of Case steamers are still chugging along over 70 years after production ended, a real testament to their skill. *J. I. Case*

most economical position for the control was called the "company notch" because of this economy. Engineers also used the expression "running hooked up" when the valve control was set for this economical position.

An important variant on the cutoff replaced the "D" type with a "piston valve," an alternative design that provided improved economy and control during the latter part of the steam traction era.

Lube Pump

Steam engine components were heavy cast-iron or steel, in constant contact with moisture. Good lubrication was essential on a steam traction engine and was provided by injecting a special oil into the inlet

NEXT PAGE
Port Huron engine wound up to redline, 250rpm. The engine uses the Woolf compound cylinder with two-stage use of steam, high pressure (the smaller portion at the front) and the low pressure portion (back).

Inside the stack is a venturi and a steam outlet called the "blower." Once you've got some pressure to work with, the fire on the grate can be blasted into fury by opening the blower valve. Then, air is sucked through the combustion chambers like a blast furnace, raising firebox temperatures and consuming fuel at a rapid rate.

steam line ahead of the valve. This oil had to be pumped in at high pressure to overcome the resistance of the steam pressure in the line, and sophisticated lube oil pumps were found on all engines to serve this function. Many steam traction engines used a "hydrostatic" lubricator. This lubricator actually used steam pressure to drive oil to the wear points on the tractor.

Governor

Governors on steam engines are curiosities to visitors to farm shows, spinning merrily at the top of the engine, but they made the operation of the engine much safer and more efficient than they would have been without it. The governor adjusted the amount of steam available to the engine and could be controlled by the engineer. That meant the engine could maintain steady speed for a threshing machine, for example, regardless of load. When a sudden load was applied to the engine—as when a couple of bundles of grain were tossed into the thresher—its natural tendency was to slow down under the load. The governor applied more pressure immediately, keeping the engine's rpm constant.

Most engines seemed to run best at 250rpm and governors normally wouldn't allow higher speeds. The governor was driven by a belt, though, and the belt sometimes broke. When that happened the engine would start to overspeed. The components of steam traction engines were designed for slow power cycles and rotational speeds; once the engine went over 300rpm or so, things started flying off—the clutch, flywheel, and sometimes the engineer. So the governor was an important device, helping to maintain safety by keeping the engine at a smooth and steady speed.

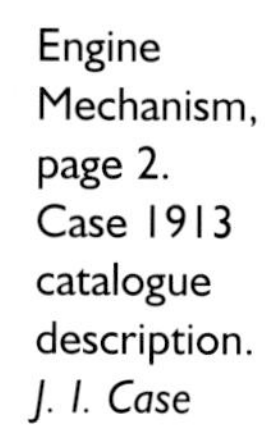

Engine Mechanism, page 2. Case 1913 catalogue description. *J. I. Case*

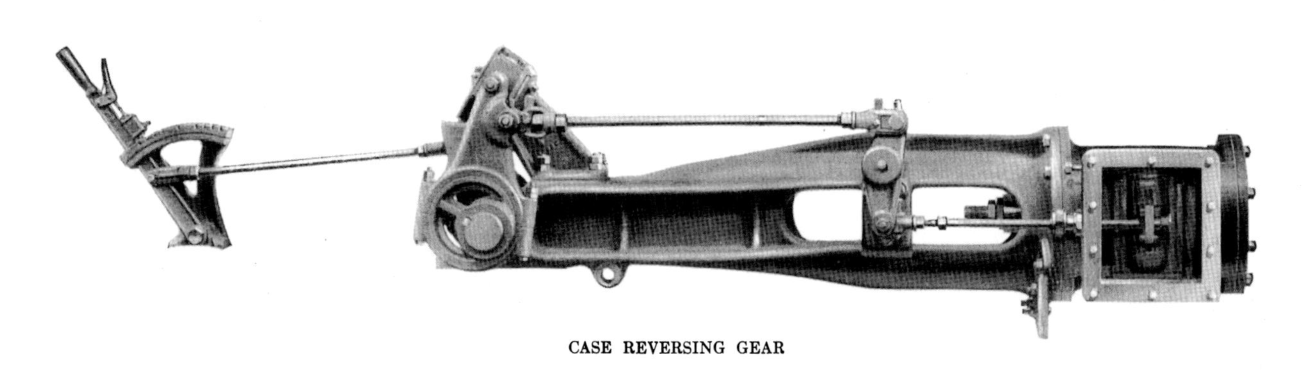

CASE REVERSING GEAR

Valve Gear. Our Valve Gear is a combination of simplicity and utility. All trappy and complex parts are avoided. Wearing parts are reduced to the smallest possible number and constructed with a view of easy adjustment. We use a single eccentric, securely fastened to the crank-shaft, to prevent slipping. The eccentric strap has an extended arm, pivoted to which is a maple block sliding in a guide. The direction of the guide may be changed by the reverse lever, and the inclination or angle at which it is set determines the direction in which the engine is to run. The degree of this angle also fixes the point of cut-off, which governs the amount of steam admitted to the cylinder during each stroke. The eccentric strap is adjustable, so that all wear can be readily taken up. The wood block is made of hard maple boiled in oil. The eccentric rod is provided with brass boxes, all wear being taken up with keys. The rocker-arm end of valve rod is also provided with brass boxes and key, which makes a valve gear on which the wear may be readily taken up at any time by the operator, thus insuring a smooth-running engine at all times. Our valve motion can be reversed quickly under a full head of steam, without danger either to piston or valve gear.

The Quadrant is provided with notches at each end, that allow the operator to adjust the movement of the valve to conform to the work the engine is doing. The quadrant is also provided with a central notch, in which position the engine will remain stationary should the throttle be opened by accident.

The Throttling Governor maintains a uniform speed of the engine at any desired rate, and can be easily adjusted. The operator is thus permitted to devote his attention elsewhere, and can depend on the governor maintaining constant and uniform speed.

The Case Oil Pump for lubricating the steam cylinder acts on the force-feed principle. It is driven by the valve gear, and is a positive feeder under all variations of temperature. When the engine is not running the pump ceases to act and no waste of expensive oil ensues.

The Fly-Wheel is of good proportion and serves as an excellent balance for the engine. The face of the fly-wheel is turned in a boring mill, with a slight crown in the middle to make the belt run true when using the engine for stationary work. It is within reach from the platform, but clear of the traction wheel, so there is no difficulty in putting on the belt should it be thrown off while the engine is running, as is the case where the fly-wheel is mounted between traction wheel and boiler.

The Friction Clutch is a most complete, efficient and convenient device for transmitting the power of the engine to the traction gearing. It is positive and reliable in its action, and can be engaged or disengaged either when the engine is at rest or when running. By means of the friction clutch the engine may be instantly disconnected from the gearing when desired for belt power. Turnbuckles with lock-nuts are provided to keep in perfect adjustment the wooden shoes which bear against the rim of the fly-wheel. With our clutch the belt may be tightened while threshing. *Without reversing the engine can be backed slowly, until the belt is as tight as desired, and without loss of valuable time.*

FLY-WHEEL AND FRICTION CLUTCH, IMPORTANT FEATURES

THE HUBER
MFG CO
THE HUBER

Firebox interior; crown sheet overhead, complete with fusible plug and numerous staybolt heads.

LEFT
Just about everything on a Huber looks a little funny, including the firebox. That has a lot to do with the Huber's "return flue" boiler design that sends combustion gasses through the water twice, once to the front, then back toward the rear, before being vented to the atmosphere.

Steam Chest

The steam chest was a large casting immediately adjacent to the cylinder that housed the valve. It was somewhat similar in function to the carburetor housing on an automobile engine—a component that contained the "fuel" just before it was fed to the power-generating cylinder. The valve slid back and forth within the steam chest.

Cylinder

If you're a gearhead like most readers of these books, you're probably used to pretty elegant engine design. By contrast, steam engine cylinders tended, though, to look somewhat like a chunk of cast-iron sewer pipe with a little better internal finish. The parts were big, heavy, and simple. Weight wasn't something manufacturers worried about when these tractors were built—durability and simplicity were the prime essentials.

Cast into the cylinder was a passage, or hole, at each end; steam came in and departed through this same port, a combined intake/exhaust port. The cylinder head was usually just a heavy cast-iron disk, bolted to the cylinder.

Piston

Steam engines have a large piston, normally about 10 inches in diameter, sliding back and forth within a mating cylinder. Steam was vented into the cylinder by a big cast-iron sliding valve. This valve admitted steam to alternating sides of the piston, pushing it first one way, then the other. Every stroke in a steam engine was a power stroke, unlike a conventional gasoline engine where every fourth stroke delivered power. The 60 horsepower Case had a 10-inch bore and 10-inch stroke. Others had much larger pistons and longer strokes.

Two piston rings formed a seal with the cylinder wall, but these rings were much heavier and thicker than those in internal combustion engines.

Flywheel

The flywheel evened out the rotation of the engine. This part of the engine was always a large, heavy casting with several functions. In addition to smoothing out the engine's cycle, the flywheel served as the pulley for belt work and the driven component of the clutch assembly.

Crosshead and Connecting Rod

A piston rod transmitted the linear motion to a crosshead, a sliding block that traveled within two (usually) V-shaped ways within the cylinder casting. This crosshead attached to a

Smokebox interior. That's the "blower" at the top of the picture, a steam line that vents to the stack. The flues get daily cleaning from this end before the engine is fired for the day.

Traction Wheels. Case 1913 catalogue description. *J. I. Case*

Traction Wheels

CONSIDERED from the view-point of traveling and hauling, a Traction Engine is no stronger than its wheels. The thousands who are using our engines will testify to the strength and durability of CASE Wheels.

The rims of CASE Traction Wheels are of steel, with malleable-iron grouters. Strong spokes of ample number are provided, all of which are inserted through the rims and securely screwed into extra large hubs.

Extension Rims for Traction Wheels. For plowing, or for use in any locality where the ground is soft or spongy, we furnish 12-inch extension rims for our 40, 50, 60, 75, 80 and 110 horse-power traction engines; and 8-inch extension rims for the 30, 40, 50, 60, 75 and 80 horse-power engines. These rims are built in the same manner and of the same material as our regular traction wheels, and are attachable or detachable in a very short space of time.

Freighting Wheels. Engines that are to be used largely for hauling purposes, particularly over rough or stony roads, should be equipped with our special freighting wheels, which we supply at extra cost. These wheels are extra strong; the hubs, spokes and tires being much heavier than our regular wheels. The spokes are thirty-two in number on the 30 horse-power; forty, on the 40, 50, 60 and 75 horse-power; and forty-eight on the 80 horse-power; and are 1-inch diameter, upset to 1⅜ inches at the threads. The tire is ¾ inch thick; the gear braces are 1¼ inches in diameter and are held to the tire by 1¼-inch rivets. In place of the malleable-iron grouters a special heavy grouter of steel is used, which has greater wearing qualities, making it especially desirable for use on stony roads. We are prepared to furnish special freighting engine wheels for 30, 40, 50, 60, 75 and 80 horse-power engines. Our 110 horse-power engines are regularly equipped with similar wheels.

Municipal Wheels. To meet the requirements of ordinances in certain municipalities which prohibit the operation of traction engines on paved streets, when equipped with regular grouters, we have designed a special traction wheel with flat grouters, which will not injure even the best paved street. These wheels are made of enormous strength to increase the weight of the engine, thereby increasing its traction power, which is somewhat lessened by the smooth grouters. We are prepared to furnish these wheels for 30, 40, 50, 60, 75, 80 and 110 horse-power engines on special orders and at an additional price.

Simplicity of Construction

IN our claims for superiority there is no one more vital to the operator than that of the simplicity of the mechanism of the engine. We have planned very carefully for the ease and convenience of the operator, by providing for an especially clear view ahead, and the facilities for easy, sharp turning. How well we have planned is illustrated in the front view of the engine. The generous size of the fire-box and the space before it, which permits of freedom in firing and operating—the working parts placed in full view and with easy access, as shown in the rear view of the engine, give to the operator the labor-saving devices which, had they been neglected, would have made his work very difficult.

With its mechanical excellence, the thoughtful addition of these small items for convenience, stamps the CASE as the engine best planned for the operator.

The pull on CASE engines is from the axle. There is no strain whatever on the boiler. The draw-bar is made with a heavy spring, as shown in the sectional cut on page 20. The tank is made from No. 12 gauge steel and is of extra capacity. The weights of the various engines are given in the specifications under each one. These figures are worthy of your consideration, being from 2,000 to 2,500 pounds less than those of other makes. If you figure your item of cost accurately, you know that you cannot afford to haul surplus tonnage in the shape of unnecessary weight of your engine.

Simplicity in design and construction, ease and convenience in operation are money-saving qualities found only in the CASE engines.

Simplicity of Construction. Case 1913 catalogue description. *J. I. Case*

Chuck Whitcher's got his Russell pretty well disassembled. That's the crankshaft and crank disk, the piston and piston rod, and the left crank bearing

connecting rod bolted to the "crank disk" or miniature flywheel on a shaft, changing the reciprocal motion to rotary. With the pedal to the metal, and 140psi in the boiler, our prototypical Case would putter along at 250rpm (redline).

Running Gear

Transmitting the engine's power from the crankshaft to the driving wheels was a substantial puzzle 125 years ago. The basic solution used a set of very simple spur gears and a large ring gear within the driving wheel, controlled by a simple clutch. Most steam tractors had one speed—slow—in forward or reverse. Some didn't have reverse. Top speed was a blazing 2.61 miles per hour for the exemplary Case 1917 tractor.

Crankshaft

Steam engines operated at rather slow speeds but produced plenty of torque at the crankshaft. That shaft would normally be at least 3 inches in diameter and would be made of the best steel the manufacturer can find. It was generally about 6 feet long and rode in carefully aligned, babbit-lined bearings. Machining this massive component to close tolerances was a good trick 100 years ago—keeping it straight was another. Keyways were cut into most crankshafts to retain gears and other drive-train components. The machining process resulted in unequal stresses in the long steel bar, which in turn resulted in slight but important out-of-true bending of the shaft, a condition corrected by skilled machinists during the process of manufacture. When Chuck Wisher had to replace the shaft on his Russell, he discovered that even modern computer-driven technology had a hard time equaling the work of those old craftsmen and their simple tools of 70 or 80 years ago.

Clutch

Nearly every steam traction engine used the same basic friction clutch design: a set of wooden shoes cammed against the inside surface of the flywheel. A simple lever control operated a series of rods and sliders that permited the engineer to gradually engage a load by applying pressure to the shoes, slowly increasing the friction, and finally matching the speeds of the engine and the driven load.

Pinion and Bull Gear and Drivetrain

Nearly all the gears on steam engines were simple cast spur gears, although some models had sophisticated bevel-geared differentials. A pinion gear on the crankshaft, engaged by the clutch assembly, provided power to an intermediate gear on typical steam tractors. An intermediate gear engaged a huge ring gear attached to the drive wheel. These gear trains accepted huge loads and seemed to have failed fairly often. But the whole drivetrain was exposed and unprotected by sissy stuff like guards or shrouds, so it was pretty easy to get at the components needing replacement.

Brakes

Few steam traction engines had any independent brake system. Instead, the engineer used the reverse control to apply steam in the opposing direction of travel. This allowed the operator to control speed on a grade—within limits, unless the gears broke. So, to drive downhill, the operator would roll forward but actually be in reverse.

Steering

A simple shaft and worm gear rotated a horizontal "windlass" shaft to provide steering control on the typical steamer. Chains attached to the front axle wrapped around this shaft, pulling on one side and releasing tension on the other. It was about as crude a concept as can be imagined, but it worked quite well. In fact, Bob Blades' 1916 Port Huron uses this system and he says, "Everybody says, 'boy, that must be tough to steer,' but it's really simple. It will turn on a dime, so much easier than the way it looks."

This large casting is the sliding component of the valve; those apertures vent inlet and exhast steam to alternate sides of the piston.

Pistons on steam engines are big, simple castings. The mass of the casting helps retain heat, inhibiting condensation.

CHAPTER FOUR

THE CARE AND FEEDING OF A STEAM TRACTOR

Dave Erb and Eldon Brumbaugh report in their excellent book on the J.I. Case company, *Full Steam Ahead,* that by 1911 two steam boilers blew up somewhere in the U.S. every day, and that in 1914, 120 people were killed, another 240 injured, in 340 explosions—during just the first six months of that year.

Steam engines contained tremendous potential power and could be very destructive when they blew. But as long as the water level was within the designed range and the safety valve was working you were quite safe. Both were easy to check and to control. Explosions happened to sloppy, complacent engineers. They also happened to people who operated steam engines that had the fusible plugs in the crown sheet welded up, or the safety valve tied down—both were somewhat common conditions long ago, when there was work that had to be done and compromises to be made. Ignorance, carelessness and laziness are always dangerous around machines.

Water for the Boiler

The quality of the boiler feed water was always a major concern for the engineer and fireman. If the farmer provided the water, or the containers to move it from the source, contamination was a real possibility. If the farmer offered a tank wagon that was used to transport milk (as was often done with the skim milk used to feed pigs) the feed water would very likely be contaminated with milk, even tiny quantities of which would cause foaming in the boiler.

A little black smoke looks good for the camera and is unavoidable when fresh coal is added to the grate, but that's wasted fuel going up the stack and a good fireman keeps it to a bare minimum. This was done for your benefit and promptly disappeared.

OPPOSITE

Look out, here comes John Tower with the family tractor. The tractor and the farm have both been in the family for many years, and the tractor still does real work to keep the farm in business. This angle shows the drive train (not OSHA approved) and front axle systems clearly.

In many localities, the fireman and engineer would have to fight the livestock for the water in the stock tank, and E.J. Murphy remembers times when the cows went thirsty for an hour or two because the stock tank was dry and all the available water was in the boiler.

Coal, Wood, Straw, or Oil

Most steam traction engines could be fired with any of these fuels, but each required the fireman to know his stuff. A coal fire must be thin and even, with no gaps on the grates and no blockages of the air passages. Firing with wood usually meant stuffing the firebox full to the top all the time. Straw was free but required the firebox be modified by the addition of some firebricks, and a special straw feeder

PAT LACEY, FIREMAN

"One of the big problems with the steam traction engine was finding competent help to run them. My dad had a fireman named Ed McNey on the 30 horsepower Avery; Ed was not only a fireman but a self-proclaimed engineer. Ed went through coal like crazy! Pa bought his coal by the carload, delivered to town, and hired a man with a wagon whose only job it was to haul the coal to where we were working, a wagon load every day. Ed went through up to 3,000 pounds of coal per day on road-building jobs! Big clouds of black smoke pouring from the stack! Safety valve popping off all the time!

"But then Ed slipped and hurt his leg and was out of action; Dad had no fireman. But Pat Lacey was standing there among the men working on the road. Pat stepped up and said, "I can fire that thing. I used to fire on the railroad." By gosh, Pa hauled him up there on the deck and put him to work—and that boy was the best fireman we ever saw! He knew just what he was doing, and he used a third less coal to do the same work!"—*Ellis Nelson*

Dr. Chuck Whitcher is Professor of Anesthesia (Emeritus) at Stanford University's Medical School; he here performs a somewhat different, but demanding, chore. The care and feeding of a steam engine requires some of the same sensitivity needed in the operating room; the steamer is a kind of organism and its health and power output respond to careful management.

was usually attached. The fireman had to continually feed straw to the fire, pushing one forkful in after another. In fact the firebox door was wide open all the time and the only thing blocking the airflow through the door was a big wad of straw, about to feed the flames.

Firing Up

The engineer could expect to sleep on the open ground, or nearby, during the annual "threshing run." He'd get up around 4 or 5 a.m. to get the boiler fired up. Most steam tractors needed two full hours or more to get up steam pressure, and no engineer wanted to have the crew standing around, ready to work, while the tractor was still simmering. In fact, it was a point of pride to be the first on a summer morning to have steam pressure—and to be able to proclaim it to the neighborhood with a long, loud blast on the whistle! One of the joys of old-time threshing was the musical salute to the sunrise from steam whistles near and far across the prairie.

The first thing a good engineer did was check the water level. Small boys were notorious for fiddling with the machinery when nobody was looking and sometimes closed the valves for the sight glass that shows the water level in the boiler. It was quite possible to have a normal level in the sight glass with low water in the boiler. So the engineer checked by opening the petcocks above and below the glass—straight water from the lower showed a normal minimum level.

The water had to be clean enough to drink—or cleaner, old-time engineers would tell you. You could use muddy creek water, or mineral-laden well water, but both caused problems. When available, some collected rainwater in cisterns, or pumped from clean ponds or lakes. If hard water or dirty water was the only thing available the engineer had to clean the boiler, scrubbing off the muddy sludge and chemically removing the lime build-up.

Then the engineer built the fire on the grate. They used oily rags, paper, kindling, and maybe some corn cobs, straw, and most added a glug or two of kerosene for insurance, then chucked in a match and the fire was lit. In a few moments, when the kindling was well established, the engineer or fireman started adding the coal.

Good coal was a precious commodity for a steam engineer. The best was "nut" coal, small lumps about the size of a goose egg.

Learning to build a good fire was a part of the steam engineer's art. The ideal was a thin layer of burning fuel evenly distributed across the grate. You didn't want it piled up or too thick or with bare patches. Too much coal cut down the airflow, disrupting the fire. An uneven layer could get the grates overheated and warp them.

The engineer stoked the fire every three to five minutes, quickly opening the door, adding a small scoop of coal, keeping the fire thin and even, and shutting the door promptly. The damper on the firebox door would allow the engineer to manage the fire, controlling the amount of air available for combustion.

One of the charming things about steam engines of all kinds was that they seemed to become living beings rather than mere chunks of metal and rubber, and as the fire got cooking, the steam engine started to come to life. The first phase was a "sweat" as all the moisture from the combustion gasses started to condense on the cold flues; it poured out of the engine from the firebox and the smokebox at each end. Soon enough the sweat ended as the metal heated; the moisture was still present but now escaped as vapor.

Filling the Oilers and Grease Cups

While the fire got to work on all that water, the junior members of the crew got to work oiling the many grease cups and oilers. This was a task repeated many times during the day, at least on good crews. Bearings were simple and wore out quickly, given the chance. Pressure oilers were installed on later steam engines, but all examples of the breed needed lots of attention from the oil can and grease cup.

The Blower

After an hour or so, the needle on the pressure gauge would come off the pin. You could hear the boiler start to simmer.

Terry Galloway supervises from on high while Chuck selects a fresh length of eucalyptus log for the firebox. The big Best tractor's steam gauge isn't off the lower peg yet although it's been fired for half an hour or more.

Long before sunrise, while most of the crowds who attend the big steam-up at Rollag, Minnesota, are still sleeping, engineers and firemen for over fifty steam traction engines clean the flues of their engines and light the fires for another day of steam-powered fun and adventure.

When there was sufficient pressure, around 15 pounds or so, the engineer could speed things up a bit by opening the vent on the firebox door and using the "blower," a vent in the stack that, through the venturi principal, sucked air through the grates and turned the fire into a blow-torch. It also blew all the accumulated soot right out the stack in a black, evil cloud. The steam pressure dropped back down, but then came back soon enough.

As the metal expanded some small "weeping" leaks would stop. The brass fittings that are threaded into steel would expand and tighten up. The engineer snugged up the hand hole cover fittings now, too.

Up to this point the boiler system had remained closed up tight. At 50 pounds pressure or so, however, the engineer would climb up on the engine and very gradually crack open the main steam valve from the steam dome. Water would have certainly accumulated in the lines. Water didn't compress and if allowed to blast into the valve and cylinder of the engine under pressure all sorts of things could have broken.

Then the engineer opened the drains on the steam chest—the large valve housing above the cylinder, and cracked the throttle a tiny bit. Water would gush from these drains at first, then sputtering steam, then plain, dry steam out of the bleeds. Only when the cylinder was fully drained and warmed could the engineer risk a full revolution of the engine.

"You wanted to keep an eye on that water level," said Stan Mayberry, "and keep about a third to a half of a glass all the time." A steam engine naturally consumed water while producing power. That water had to be replaced, the upper level kept about in the middle of the sight glass, and monitoring that glass was a frequent activity for the engineer. Sooner or later, though, the level would drop and more would have to be added. One problem, though—how did you get the water in? It couldn't be poured in because the steam pressure would escape and the whole system would have to be

Once some pressure is available to work with, you open the petcock drains—most steam traction engines have a single control that opens or closes all of them at once—to warm the lines, valve, cylinder, and piston. Lots of water will have condensed and that all needs to be purged. Glen Christoffersen and Chuck Whitcher inspect the works as the monstrous old Best comes alive again. And—don't worry, those coveralls will be authentically filthy by the end of the day.

pressurized again. Instead, water was pumped in under extreme high pressure by an injector or mechanical pump. The boiler feed passed through a heat exchanger before going into the boiler, bringing it up to near the temperature of the water in the boiler.

That water may have come from a creek, cistern, horse trough, or well, but most of the time the steam engine would be working far from a source that permitted direct pumping. Instead, the steam tractor normally had a tender nearby with a 350 gallon tank of good boiler feed water and a few hundred pounds of coal.

On the Road Again

Once the pressure was up, getting the tractor underway was surprisingly simple. Since most steam traction engines had only one gear for travel, choices were limited to forward or reverse. Here's the sequence on Bob Blade's 1916 Port Huron:

First, make sure the gauge reads 135 pounds Then gently position the valve control on the "Johnson Bar" to the front for forward motion or toward the back to go in reverse. Next, engage the clutch, and open the petcocks to drain any water in the cylinders or lines. Crack the throttle and the tractor will start to lumber forward (or backward). Finally, close the petcocks and off you go, at about 2miles per hour!

Downhill Danger

The vulnerable crown sheet at the rear of the boiler could easily be exposed while the tractor was traveling downhill so engineers took special precautions in hilly terrain. "If you were going to be driving down a long hill, you'd want to let your fire burn down pretty low," Stan Mayberry says, "so it didn't get too hot on the crown sheet and burn your fusible plug out."

Some designs sloped the crown sheet toward the back, anticipating that

Bob Blades sets the cutoff valve on his nifty Port Huron. The two wires visible in the shot allow the governor to be adjusted "on the fly," within a range of about 100rpm.

down-hill water level. But mostly it was up to the engineer to be ready for hills—in both directions. With some tractors, exposure of the crown sheet was quite likely on a slope unless a lot of water was carried in the boiler. But that was a problem, too, because of the risk of "priming," or getting liquid water into the steam line, causing knocking and risking damage to the piston and cylinder.

Anatomy of an Explosion

Explosions were once fairly common and *almost* always preventable. Despite the suspicions of steam tractor engineers, these disasters were seldom due to faulty construction; nearly all occurred when the engineer was negligent, one way or another.

The blast, when it happened, wasn't a complete surprise to the victims. You could be chugging along, with the pressure gauge indicating a nice, normal reading. Open the fire door to toss in a bit more coal, though, and you might have seen the red hot crown sheet reflected in the polished surface of the shovel. When that happened—and it did-you had a few precious seconds before somebody could collect your life insurance policy.

It was easy enough for the water level to fall below the level of the crown sheet, even on level terrain; it would certainly happen going down a steep hill. If it was only momentary, if the water was sloshing around and keeping the steel wet, you were safe and sound.

But when that crown sheet was exposed long enough to start getting red hot, the metal would begin to soften and expand. The crown sheet would start to push and pull against the staybolts and braces. If it got very hot, and very soft, the steam pressure would distort it. If it got very hot and water sloshed across it, the result was "flash steam," with pressure estimated at between 600psi and 1000psi. That would often start a rupture.

It didn't take much of a failure to start the explosion. And the explosion was a slow, violent one. Remember, the water in that boiler wasn't at 212 degrees Fahrenheit but well over 300 degrees; the pressure in the boiler was the only thing that kept it liquid. Once the pressure started to drop—either by a sudden opening of the throttle (which vented pressure) or by a crack in the boiler that provided a vent to the atmosphere, the liquid water would start to form steam. The pressure would suddenly rocket upwards, expanding the failure in the crown sheet. The small hole would very quickly expand, venting more steam, and lowering the pressure in the boiler. All that 350 degree water was now exposed to atmospheric pressure and it rapidly turned to steam. Remember that one cubic inch of water evaporated to a cubic foot of steam at sea level air pressure—an expansion 1,728 times the original volume.

But the boiler didn't hold a cubic inch of water—it held hundreds of gallons of water, each expanding over 1,700 times. It happened within a second or two, with

Sidemount engines are certainly the most common configuration but you need to be a mountain goat to attend to their wants and needs. The upper green cylinder houses the piston and is the foundation for the engine. The long lower green tube pre-heats boiler feed water before it is injected into the boiler; without it the fresh feed water would chill the boiler.

a tremendous rush and whoosh, rather than a sharp bang.

A huge cloud of gray soot and steam suddenly roared out of the tractor, usually from the firebox. The fire was instantly extinguished, but it was now too late. The thick plate steel of the firebox and surrounding area was ripped apart like wet cardboard.

The tractor, no matter how big, would probably become airborne. Some have been reported to travel hundreds of yards. It could rocket along on its wheels, going a lot faster than the 2.5 miles per hour the engine and geartrain normally provided. Depending on just where the steam vented, the tractor might have flown end-over-end before coming back to earth.

The engineer might have been thrown hundreds of feet in the air. And it was a lucky engineer who was dead on impact, because the burns from this superheated steam could be fearful and lingering. Newspaper reports from the early part of the century often contained reports about such accidents and the names of those who lingered a day or two in agony before expiring.

Fusible Plug

Boiler explosions shouldn't happen, even with the most negligent engineer. It didn't take early steam engine designers long to realize that some kind of safety device was required. The safety valve was one solution, and it worked well, most of the time. It was really just a heavy spring-loaded valve that would release at a preset pressure. But these valves corroded, stuck, or got coated with mineral deposits from the boiler feed water. Sometimes they were adjusted improperly. Sometimes the engineer tied them down. But they worked pretty well, and few boiler explosions resulted from faulty safety valves. The normal culprit was a sudden flash of steam from a red hot crown sheet that came in contact with water.

So the designers came up with something that the boneheaded engineer wouldn't easily defeat, the fusible plug. This device was a brass plug with a lead or babbit center, screwed into a fitting in the crown sheet above the firebox. As long as water was in contact with the plug and the crown sheet, the babbit would remain solid. But as soon as the crown sheet got dry, and as long as there was a good fire on the grates, that babbit core would melt at a pretty low temperature. When that happened, all the steam in the boiler vented right into the firebox. The steam immediately extinguished the fire. The pressure dropped to atmospheric levels. The engineer sweared a blue streak, looked embarrassed, and made excuses. But no other damage was done, and the boiler would soon be back in business.

Well, if this plug worked so well, what about those two explosions per day that were once the norm? Those boilers had fusible plugs, too, in most cases. But if a boiler wasn't cleaned regularly, particularly if the scale and lime buildup wasn't removed, a hard crust could form over the plug. When the babbit melted out, nothing was vented because the hole was blocked by scale. But the worst cause, and the most unforgivable, resulted when the engineer replaced the fusible plug with a solid one. It happened all too often.

Death and Disaster

While boiler explosions seemed like the biggest hazard to a steam engineer and fireman, most were hurt and killed crossing bridges. If two steam traction engines exploded every day, 80 years ago, then it seemed ten tractors fell through bridges. The carnage was quite considerable.

Most professional threshing rig operators planned a "run" to avoid crossing streams or terrain that included a bridge, unless that bridge was extremely trustworthy. Few of them in rural areas were. The Brotherhood of Threshermen, a professional organization, and "American Thresherman" magazine both lobbied state and federal governments to improve the quality of rural bridges, with some localized success.

Terry Kubicek's Czech ancestors settled in Nebraska and farmed with steam, back around the turn of the century. Terry maintains the family tradition with three steamers of his own, one shared with a brother, and his three-year-old daughter even has one of her own. Terry's grandfather and great-grandfather helped enact an important Nebraska law with an anecdote and a lawsuit about a steam traction engine they'd just bought, back around World War I. Terry tells how it happened:

All Steamed Up And No Place To Go
by Terry Kubicek

"Great grandfather and grandfather, who both lived near Crete, Nebraska bought a steam engine. The sales office was located in Seward, Nebraska some 20 miles away. The terms of sale were essentially payment due on delivery to the farm, which was about 3 miles south of Crete."

"The steam engine was loaded on the Chicago, Burlington, and Quincy Railroad at Seward and duly delivered to the depot at Crete. From Crete the engine was to be fired up and driven to the farm. Traveling at 1.5 to 2.0 miles per hour plus a stop for fuel and water, progress was slow but steady; the engine was in the capable hands of the sales company representative".

"All went well until the engine fell through a bridge, killing the capable sales company representative. The company, not to be deterred or detoured, organized a salvage attempt and pulled the steam engine out of the creek with a number of teams of horses from local farmers. At that point payment was demanded and refused as the engine was not yet delivered at the farm and had not been refired."

"Great grandfather Matjej/grandfather Charles required that it be fired up. Much to the chagrin of the company, it would not hold steam and so payment was refused. The company, intent on the sale, purportedly made repairs but without firing the boiler dragged the steam engine to the farm and once again demanded payment."

"Once again, the skeptical Czech farmers in the persona of my great grandfather/grandfather refused payment and decided to fire up the boiler on their own. Not surprisingly the

There go the R&R boys, Sonny Rowlands and Orman Rawlings, and their big toy, playing in the dirt again. Their Advance is fueled by LPG so you won't see any black smoke coming from the stack. Their helping with spring planting at the annual antique farming "play day" at the Corona Ranch near Temecula, California, held in April.

"There were a lot more single cylinder engines than any other types, but the double cylinder types gave smoother power and there was no "dead center" problem to contend with. My dad preferred the double cylinder engines, but he did a lot of plowing and road work and thrashing. But for all the plowing he did, he found that under-mount 30 horsepower Avery without any equal. The boiler was good to 150psi, but you couldn't tell much difference between the 30 horsepower and the 40 horsepower Avery running 200 pounds. I used that 30 horsepower to thrash in 1925—we pulled the biggest thrasher Avery made with it, without a problem."—*Ellis Nelson*

1906
16 HORSE POWER

boiler did not hold steam and payment was again refused."

"The company eventually retrieved their steam engine and sued for the purchase price. The case went to the Nebraska Supreme Court in 1913-14 and the Court held for the defendants in an opinion stating that a machine must be in satisfactory operating condition for the purpose intended by the parties and if a material aspect of the operations is not as intended by the purchaser and so advertised by the seller then no payment should be reasonably expected and none due. In essence, payment on delivery meant that the machine had to be in good working order before demand for payment would be actionable.

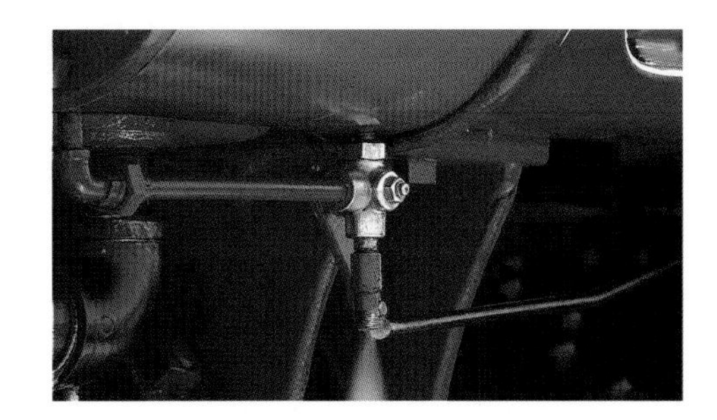

That steam is venting from the bottom of the cylinder, one line from each side of the piston. A rod system operates the control from the relative comfort and convenience of the cab and is linked to all the other drains on the engine. Liquid water in the cylinder causes "pounding"—enough of it at the wrong moment can smash the heavy steel castings into junk.

Eighty-three years later, that case is still good contract law in Nebraska."

"At some point another steam engine was purchased and operated for a number of years with neighbors and relatives. Eventually grandfather Charles and his brother Fredrick owned an engine together, but Fred did not like to do the pre-season/post-season maintenance and Charles bought out his interest. Grandfather Charles's steam engine survived until the early 1950s and although field ready, was cut up for scrap for $50.00."

"Grandfather Charles wanted to buy some other piece of equipment and my father Lumir let it happen. Even then, the cutting was done

This big J. I. Case is one of the performers at the excellent show held every September at Booneville, Missouri. Case outsold the competition by about 3 to 1, building over 38,000 steam traction engines

The Booneville, Missouri, show isn't nearly as big as Rollag, but you might not have to drive as far, either—it's along Interstate 70 between St. Louis and Kansas City. A half dozen steam traction engines usually show up and you can visit with the owners all you want. They might give you some driving lessons and if you ask real nice, let you help with the greasing and coal-hauling.

by my father's first cousin Adolph. The whistle is the only surviving part, and is now on the replacement engine. For ten years, I looked for that replacement machine and finally at the Camp Creek Threshers Swap Meet in 1993 purchased the machine."

"My wife and I currently own three traction engines–a Nichols and Shepard, a Keck-Gonnerman, a Minneapolis Threshing Machine Company as well as a Geiser-Peerless portable. My brother and I own a Case, and my daughter, who is probably the youngest steam engineer in Nebraska at age three, owns a shop built upright steam boiler on a model A frame, with 4 speed gear box and twin cylinder steam chest. Believe me, she is cute as a button, and hell on wheels when she gets steamed up. The home-built is pretty nice too."

Testing and Certification

Good sense (and some state regulations) require operators of steam engines to test and sometimes have certified boiler systems. Even so, most present day owners of steam tractors are very careful about the condition of their boilers and associated components. Few old steam tractors, sitting outside for a few decades, would have its boiler plates as thick and strong as when new.

Modern technology helps the restorer in some very reassuring ways. One is the use of portable ultra-sound diagnostic instruments to examine the boiler shell for flaws—corrosion, which is common, and cracks, which are rare. "A lot of guys won't buy a steamer without an ultra-sound test," Stan Mayberry says, "because it lets you test the thickness of the boiler barrel and other components, to see what kind of shape the thing is in."

The hydrostatic test is another verification technique, sometimes required on a regular basis by state regulators. This test involves filling the boiler full of water—which isn't compressible—and applying pressure with a simple hand pump. A boiler rated at, say, 150psi will be pressurized to 200psi, well over its normal maximum. Then everybody waits and watches for leaks. Often a little water will weep from a staybolt fitting, a plug, or elsewhere on the boiler or the engine, and is easily repaired.

Coming Down With the Flues

You'd think boiler flues would be difficult to find now that steam locomotives and tractors are pretty much history, but there are plenty of places to buy new ones. In fact, steam boilers are still pretty common in factories and elsewhere, for powering steam engines and for providing steam heat. There are dealers, too, who specialize in supplying restorers of tractors and locomotives with flues and related materials; there are still many hundreds of each in operation—enough to keep support businesses busy.

ILLINOIS THRESHER COMPANY
ILLINOIS THRESHER COMPANY
SYCAMORE ILLINOIS U.S.A.

CHAPTER FIVE

THRESHING TIME

A primary use for the steam traction engine on the farm was the annual ritual of threshing—or "thrashing" as it was always pronounced—the small grain crop. The steamer was a godsend when it arrived after the Civil War, making possible the rapid, efficient, economical harvest of the crop. Since part of that crop, the acres of oats grown on most farms, would feed the work horses for a whole year, the importance of this crop was crucial to the success of the farm.

Steam engines and threshing machines were too expensive and too complicated for most nineteenth century farmers to buy for personal use. Instead, a group of neighbors sometimes formed a group called a "threshing ring" to share in the costs of equipment as well as the labor required. Since the size of the farms, and the amount of grain to be threshed, varied from one farm in the ring to the next, systems for equalizing the investment of each participant were developed. These could be quite elaborate and complex or informal. Threshing rings in Holmes County, Ohio, survived until very recently and still exist among some Amish neighbors.

A Minneapolis engine begins its day, still purging liquid water, with the first slow cycles of the huge 10-inch piston.

OPPOSITE

The chore of picking up the grain bundles was normally done by teams of horses and wagons, but this Illinois Thresher Company steam traction engine is doing the chore at Waverly, Nebraska on a bright July day.

Custom or Contract Threshing

An alternative to the threshing ring, with all its potential for conflict among neighbors, was the custom thresher. Lyle Hoffmaster and Ellis Nelson both did custom threshing during the 1930s, and Lyle's dad and grandfather did the same work before him, all with steam power.

There were some real advantages to this system. Once a farmer got his grain cut with the binder and in the shock or in a big stack, it was pretty safe from the weather. The threshing could wait until it was convenient, even in the middle of winter, and Lyle's dad even threshed with his steam traction engine and separator in February. Today, you've got to combine grain before the heads shatter, normally a brief period in early summer.

John Hoffmaster was just one of 1,000s of men who owned and operated steam traction engines primarily for use with a separator. John's business was with the farms of central Nebraska, back in the 1920s. He provided the engine, the separator, and the crew of three to seven men and boys. The farmer normally provided another five to seven men to drive the bundle wagons and haul the grain to town.

A Day In the Life of a Custom Thresherman

For the engineer, a normal day began about 5 a.m. That was when the fire needed to be started to have sufficient steam up for the beginning of the work day at 7 a.m. The engineer could earn ten

Stan Mayberry engages the clutch that puts the big flywheel into motion. The belt on that pulley turns a large Case thresher and can help thresh up to 2,000 bushels of grain in a long summer day.

dollars a day, but he *earned* that princely sum. Even before the fire was warming the cold metal of the boiler, the each of the tubes had to be carefully cleaned. The coal and water bunkers needed to be filled, once in the morning and again at noon under normal conditions and sometimes more often.

It was the host farmer's responsibility to provide meals for the crew, but breakfast was sometimes more than the ladies of the house could manage. Lyle Hoffmaster recalls that his father's crews sometimes had to wait till dinner at noon to be fed. While the engineer scurried around making repairs and getting up steam, the rest of the crew would roll out of their nests. In the early days, before the 1920s, it was common for the crew to sleep at the farm where the day's work had been done. Sometimes, but not often, a bed was provided inside the house; more often, you brought your own bedroll and spread it under the stars in summer or in the barn's haymow during cooler weather.

The fireman might be a young lad or an older man, but he needed to be pretty fit. The coal might have arrived in big chucks and needed to be broken into fist-sized pieces for efficient firing. That was work!

If the crew was lucky, and it usually was, breakfast would be on the tables sometime around 6:30 A.M. The menu depended a lot on the ethnic background of the host farmer, but oatmeal, pancakes, biscuits, eggs, bacon, and gallons of coffee were standard fare. German and Swedish farm families were likely to produce exotic sausage, too, and fried potatoes by the platterful.

But by 7 a.m., the first bundles should be tossed in the feeder, and the steamer chugged along in a businesslike way. If everybody knew their business and nothing broke, the engine and separator could hum along all morning, cleaning grain at the rate of 100 or more bushels an hour sometimes.

PLOWING WITH STEAM

by Ellis Nelson

Plowing with the big steam traction engines was wonderful —if you were man enough to handle it! I am pretty much prejudiced about this, but there was only one real plow engine ever made and that was the Reeves. I still have an early Reeves 25 horsepower cross-compound plow engine sold around Lebec, Kansas. There was a tremendous difference between running an steam engine for thrashing and for using one to plow. A man who could do a good job with a thrasher might not be worth his salt with the plow.

The steam traction plowing engine wasn't a choice for families trying to farm out on the prairie, it was forced upon them. You couldn't break the prairie with horses, not in any businesslike way. It took four to six head of horses to pull a single bottom plow in that virgin prairie sod. The heat was terrible and the flies not much better. Oxen were tried, but they would take off for a pond or the shade of a tree, and the hell with you! You could kill an ox before he would move if he didn't want to pull. Breaking 10 to 20 acres in a year was considered good progress, until the steam traction engine arrived.

So you had to have a steam engine! But a lot of those that bought them couldn't handle the steam traction engines. It wasn't something they were accustomed to, and a lot didn't adapt to the new ways. So farmers hired engineers and people with tractors to do their plowing for them.

The steam engine technology swept over that country like a wave. The western parts of Kansas, Oklahoma, the Red River Valley, and some other places were finally open around the turn of the century. This was vast country, home to the "bonanza" farms where a tractor pulling 14 bottoms would plow in one direction until noon, turn around and plow all the way back. It was done with horses first, and it killed a lot of horses. The steamer was a natural for this work, and in the west there was no choice. But the old-timers hated the steam engine!

It was about 1904, I think, that you could buy the big ganged engine plows, and some were the ultimate in mechanical ingenuity. With the Avery plow, when you pulled the trip rope, every plow came up out of the ground at the same point —a perfectly straight line! You could buy these ganged plows with from four to 16 bottoms.

Most everybody would stop plowing to take on water or coal, but the more progressive did it on the "fly." The water-hauler left a full tank where the engineer could drop the empty and hook up the full tank. Then the water hose was dropped in, and the tank trailed alongside, attached to an auxiliary hitch we called a "crane." That made for real plowing efficiency.

Some operators wouldn't bother with repairs. The efficient operators had a blacksmith around to sharpen the shares and repairing parts on the engine.

At noon a blast on the whistle would summon the teamsters in from the fields and the grain bin for dinner. The engineer and fireman stripped off the grimy overalls they wore over their work clothes, joined the other men at the wash pans to scrub up, then parked themselves at the two big tables on the grass outside the kitchen. Few farm tables could accommodate the twelve or fourteen men who helped with the threshing, so boards and sawhorses normally fit the bill.

Eat Like A Thresherman

"It's like Thanksgiving every day!" John Hoffmaster used to say about the dinners offered by farm wives to the crews of harvest hands, and he spoke for a multitude. The competition between farm women for the title of Best Cook in the neighborhood was a part of the legend and lore of the American family farm. Most farm women worked longer and maybe harder than the threshers themselves to prepare the noon meal. While a roast beef and perhaps a ham might hold center stage, some added fried chicken by the bushel, and heaps of potatoes were common. Ice cream sometimes appeared, but not often in the days before refrigeration.

But not quite every meal was such a feast. Lyle Hoffmaster tells of the time his dad took his crew in for dinner at a place where the logistics seemed to have gotten the better of the cook. A crew of hungry threshermen sat down to a dinner of boiled, over-salted potatoes, and nothing more. No butter, sour cream, or anything to enliven them, just plain boiled potatoes. But there was a dessert—pie, in several flavors, and that's what fueled the crew for the rest of the day.

A Day's Work

A toot on the whistle about 1 p.m. signaled the beginning of the afternoon's labors. The engineer and fireman would probably get back to work before the rest of the men. They would insure the water tank was refilled and the coal supply replenished and all the oilers and grease cups topped off before the beginning of the afternoon's work. The boiler would have been allowed to cool and simmer during the noon hour; now it would have needed to be recharged, the blower opened, and steam brought back up to working pressure.

By now the whole crew would have had a good idea about the job—if it could, or couldn't be finished in one day. The typical farm only needed one day's attention from a good engine and separator crew, but a big job could last for days.

The last bundle wagons came in about a quarter till six, if the rig was to stay for a second day. Those bundles waited till morning and were the first through. The fire was allowed to slowly die out; the boiler gradually cooled.

A supper would probably await the tired threshers—most likely left-overs

ANCE
MELY

You'd smile too if you owned such a sleek, sturdy steam traction engine. Stan Mayberry and the Advance are star performers every year at the delightful little show at Hamilton, Missouri. Such small local shows are probably the best way to see steam traction engines in action. The shows aren't always promoted heavily so you've got to look for them, but it's worth it; small crowds, low prices for the gate and the pork tenderloin sandwiches and home-brewed lemonade. And you can chat with Stan, too, at no extra charge.

"In 1919, my dad and W D Parker had a whole pile of land up near West Oak, North Dakota. They plowed about 900 acres of land that spring with that 'undermount' Avery, pulling 14 bottoms! Every Saturday and Sunday I'd be out there, watching. The engineer, old Bill Balance, he liked to see me coming! I was 11 years old, and he let me climb up in the cab with him and steer. Then he'd get off and run along side, listening to the machinery and watching how everything was working. I could hold it in the furrow pretty good, even then."—
Ellis Nelson

A Keck-Gonnerman all fired up and hot to trot at Adrian, Missouri.

from the noon meal, but sometimes another elaborate feast. The cooks would be cleaning up till nearly midnight, regardless.

Moving On

The next job was probably on the next farm, or one nearby, and the move may have been made either in the late afternoon or early the next morning. Either way, the tractor and separator would try to go across country, rather than by road, to the next setup. This required having the farmer cut his fence. Some may have been reluctant, but to decline was considered very poor manners.

Shutting Down and Putting Away

One great way to destroy a steam engine was to use it for a day or two, then park it and walk away. This was pretty much guaranteed to turn your big, rare, powerful steam tractor into a pile of scrap metal in days or, at the most, months. A freeze would expand any water in the boiler and other components, breaking even the strongest cylinder and heaviest boiler plate steel. Acids left in the ash from burned coal or wood would corrode the bottom of the firebox and the grates. Rain or snow would fall down an uncovered stack, rusting the floor of the smokebox.

"When we got done with the tractor—in the fall, back in the old days,

AN OLD-TIME THRASHERMAN

Lyle Hoffmaster

My dad was born in 1887, and my grand-dad was a "thrasherman" who bought one of the first Case steam engines in 1879.

Grand-dad used that portable engine for thrashing for years, then traded it on a new one. He owned seven or eight over the years and never wore a one out. That portable Case went to a farm about 4 miles north of York, Nebraska—where it blew up! The marks are still on the outside of that house, where it was hit by pieces of metal from Grand-dad's 1879 Case engine.

The second engine was another Case, a center-crank model with a straight flue. That was followed by a center-crank with a return-flue boiler, and that's the one my father started on. But Grandfather could see that there were better engines, and the next one was the only steam traction engine our family has ever named: a 13 horsepower Nichols & Shepherd known to us as Old Betsy. It was a very good engine, but at only 13 horsepower it was somewhat underpowered for the kinds of thrashers then coming on the market.

They had some problems with some boilers, and it almost broke Nichols & Shepherd because they tried to fix every one, and it cost them a bundle. But when they came out with a new high-pressure boiler, my grandfather took them at their word, and bought one. It carried 175 pounds of steam. When Dad was thrashing with it, early in the morning when the dew was still on the grain and it was pulling hard, the men could hear him at the grain elevator at Bennedict two and a half miles away; they knew when he started and they knew when to expect the grain to start coming in from that rig!

His last outfit was a 15 horsepower Case, purchased new in 1907. Grandfather probably bought too small an outfit for the work to be done, but we had bridges in that country and Nebraska didn't have enough timber to build the bridges very sturdy, with cottonwood planking. Dad would sometimes have the drive wheels go through the planking, leaving the steamer suspended on the bridge stringers. He was such an artist with that steam engine that he could chain a railroad tie to the wheels, then apply power to bring them around and get the tractor up again, and stop before the tie came back to hit him on the tailboard. That takes exquisite control with a single-cylinder sidemount engine!

The thing that made Case so successful wasn't that their machinery was all that much better—it wasn't! The thing that sold the Case steam tractors was their credit policy. Farmers were always short of cash, but they could get a Case machine when they needed it. Their grain didn't go to waste. So Grandfather was quite a Case man!

My dad knew steam almost from the day he was a baby. A horse fell on him as a child, injuring a leg, so all he did after that was work on the steam engine.

I grew up out in central Nebraska—good level land, and most everybody raised small grains. My dad had a custom thrashing outfit and ran between 100 and 125 days each year, powered by a steam traction engine. I learned to do repairs on the thrashers and by the time I was in my late teens I could do just about anything on them. When I was a kid I wasn't interested in sports or anything except machines; my mother was sending off for steam engine catalogues for me before I was in first grade. I could forge-weld by the time I was sixteen, and run a lathe before that. I became expert at replacing the poured babbit bearings used in thrashers and steam engines before I was 20. I'm a bit too young to have seen steam at its height, but I have had the pleasure of seeing two new steam engines in the dealer's shop window back when I was a kid in the 1920s.

Later, I was the youngest member of the Illinois Brotherhood of Thrashermen to own his own machine, and I thrashed for 11 years, and helped pay my way through college with that thrashing outfit.

after a show today—we drained all the water and washed out the boiler real well. We pulled the hand-hole plates to allow the air to circulate, helping dry out the inside of the boiler. All the ashes had to be cleaned out—they hold moisture and chemicals in the ash combine with water to form a corrosive that will eat up steel. Then you needed to use some 'steam cylinder lubricant' in the valve and cylinder. Normally that was injected in the steam line while the engine was running, but you need to add some more if the tractor was going to set for a while. It was a good idea to add a little boiler compound every year, to eliminate some of the scale and lime build-up," said Ellis Nelson.

All the valves and water lines had to be drained to prevent water damage. The nuts on all the fittings with packing got loosened up, too, permitting expansion when freezing weather arrived. The engineer would normally try to pump the oiler several times just before the engine was shut down, preventing rust during down time.

Finally, the tractor was supported on blocks, up off the ground, then—with all the innards clean and dry—the hand hole covers got replaced, the stack got covered with an old can, and all exterior openings plugged. "You don't want 'mud-daubers' or mice getting in your boiler over the winter, making nests. That makes for a mess, and I have had the problem!" Stan Mayberry says.

NEXT PAGE
The steam traction engine was one of the factors that made American prosperity possible. With the threshing machine, with big acreages, with efficient grain plowing and planting, a family farm could grow huge quantities of small grains. Rigs like this one could, and often did, thresh and send to market 1,500 to 2,000 bushels of wheat, barley, or oats from a single farm in a single day.

ASE

John Tower's Advance Rumley steamer is a working machine that still does the job it was designed for almost 100 years ago. John is demonstrating the tractor at a show here, but he fires it up to power a feed mill at his family's old "home place" when there's work to be done.

"I had the problem, too!" Glen Christoffersen said, but his problem was much worse. "We had some rats that got inside the Best boiler somehow. They were still there, we discovered, after we filled it and fired it up. Man, did that boiler foam! And the steam had the worst smell imaginable! It wasn't really apparent until we were hauling people around on the hay racks—but then they started jumping off in a hurry! Ever since, we put wire screens over the hand-holes when we put the thing away."

Boiler Feed Water

Engineers tried all kinds of tricks to prevent liming and scale build-up in the boiler. Some added potato peelings,willow bark, white oak bark, or "sody ash," to acidify the water a bit, neutralizing the minerals normally present in well water.

Bob Blade's Port Huron

Bob Blades and his son Dane are star performers at several northern Missouri farm shows every year, threshing and sawing with the family's 19-65 1916 Port Huron, serial number 7738.

Young Dane represents the new generation of "steam fiends" and is a fully qualified, experienced, and knowledgeable engineer at the age of 16 years old. That's a bit younger than was typical for a lad back when the Port Huron was new, but by 15 or, 16, many boys would be learning the ropes on their dad's engines.

The Blade's Port Huron came from the central Iowa town of Leon and was in good shape when purchased. "I've always thought it was a 1916 model," Bob says, "because the rear wheels have

This is what the steamer was born and bred to do, a service it performed beautifully and in huge numbers for about a half century—from about 1890 to the beginning of World War II. That prosperous farm on the end of the belt was pretty much bought and paid for through the efforts of the steam engine. Without it, the American farm would have largely remained a small, hard scrabble, inefficient operation. *J.I Case*

casting dates in them, both in December of 1915. The tractor would have been assembled the next year, so that's all I have to go on. We ran it the day we brought it home and have used it on my sawmill quite a bit, so it still is a working machine. We repainted and re-flued it, replaced the axles and put in some new bearings during that first winter, but it was in working condition when I got it."

The Port Huron uses a compound engine of the Wolfe design, one long piston rod with two cylinders on the same axis. It was a very popular design and a standard feature on most Port Hurons. Bob's 19-65 uses a 6.75-inch and a 9.5-inch piston, both with a 9 inch stroke.

"When the 19-65 was new it was rated at 175 pounds but I run it now at about 135 pounds; at that pressure I am probably getting about 15 horsepower at the drawbar and maybe 50 horsepower at the flywheel."

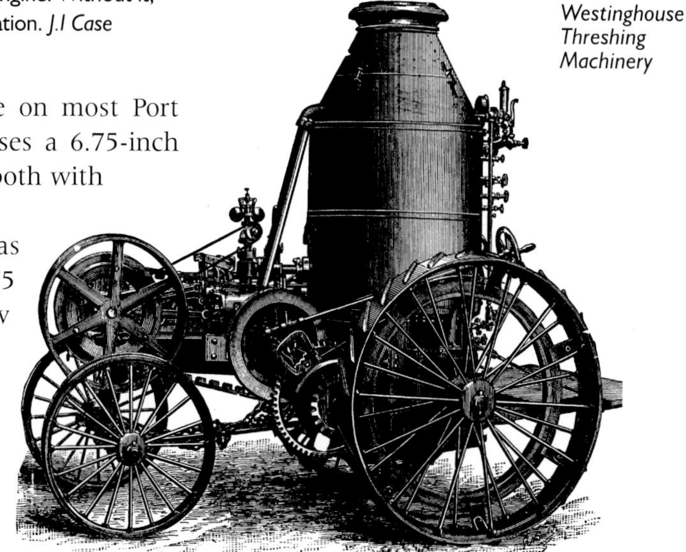

Westinghouse Threshing Machinery

THE
BEST
MANFG. CO.
SAN LEANDRO

CHAPTER SIX

PRESERVING THE PAST

Profile: Ardenwood's "Best"

Not every steam tractor went to work on the farm, although most of the ones still around are examples of that sturdy breed. One of the most unusual preserved steam tractors is the very handsome and quite famous Best Steam Traction Engine 185, an example of the large western tractor industry that once flourished in California.

The Best on display today at Fremont, California's Ardenwood Historic Farm is owned by the Oakland Museum and looks much as it did when new, back in 1904. It is a giant in every respect, 17 feet tall, with wheels 8 feet in diameter. It is rated at 110 horsepower, a tremendous figure for the time; top speed is only 3.5 miles per hour but it will pull pretty much anything you can hitch to it at that speed.

The Best Manufacturing Company is one of the ancestors of today's Caterpillar Corporation, and occupied a factory site on Davis Street in San Leandro, California, when the tractor was built (and where a Caterpillar Plant is today). Best was a successful, innovative, tractor and combine builder around the turn of the last century, and a keen competitor with the Holt Company until 1925, the year both companies merged and the Caterpillar company was born.

Terry Galloway is the great-grandson of Daniel Best, inventor and manufacturer of some of the earliest grain combines, tractors, and harvesting machinery. Terry is at the helm of the beautifully restored Best No. 185 110hp steam traction engine he helped find and restore to grandeur.

OPPOSITE
At 17ft to the top of the stack, the Best isn't really suitable for most farm operations. You certainly couldn't fit it in most barns or machine sheds. Its career was spent in the woods, moving logs and powering a sawmill. That was a common chore for a great many steam traction engines, a profession they generally performed with great success.

The tractor sold to the California Sugar and White Pine Lumber Company for about $4,500, another giant figure for that time, and went to work on the steep slopes of Mt. Lassen. After 18 years of faithful service No. 185 was sold to the Collins Ritts Manufacturing Company for $750. Ritts used it in a lumber mill, a common use for steam engines, then sold it again a few years later, this time for $500. This time old No. 185 was retired from active service; it was parked in front of a resort as a tourist attraction.

During the late 1960s, the Oakland Museum started looking for a good example of the Best steam tractors built by the Best company during their years in the city, 1889 to 1912. A search began and a fund established. Only 300 or so were ever built, many for farm use, and very few survived.

One of the key players in the search was Terry Galloway, the great-grandson of Daniel Best, the founder of the company (and one of the most influential designers of agricultural equipment of his day). Terry says, "We started out searching all the known Best tractors, and all we had seen of these machines

On the road again, the Oakland Museum's glittering, beautifully restored 110hp Best ambles off to a day's duty at Ardenwood Historic Farm near Fremont, California. Terry Galloway, great-grandson of manufacturer Daniel Best, is at the helm.

were archival photographs. When we first saw a real one—that was an experience I will never forget! That first one was up in the Gold Rush country, at Murphys, California. We considered each on several levels: Was it available? Could it be restored to operating condition?"

The search team found very few survivors. None were really available or readily restorable. But No. 185 was found at Shingletown, California, and the owner was willing to consider offers, at least. An old-time Best steam tractor engineer, Art Phillips, was driven to take a look at the machine. Art walked around the old steamer, opened the firebox door—then slammed it, hard. He listened to the ringing sound coming from the old boiler plate and steel flues. "You could fire it up right here and drive it up on your flatbed!" Art proclaimed. "All the tubes are tight, they're in good shape, and there are no holes in it."

Well, parts of it were in good shape, but a tree had fallen across the tractor and crushed some of the plumbing and other components. A deal was cut with the owner, all the bronze parts removed for safe keeping, and some foundation money acquired for acquisition and restoration. A Caterpillar dealer, Peterson Tractor in San Leandro, offered their best driver and biggest truck and lowboy combination. The old Best steamer was finally loaded and the truck started down the mountain, back to San Leandro where No. 185 was born over 60 years before.

The road back was a twisting, narrow little mountain road and the huge tractor, even without its smoke stack and on a lowboy trailer, swayed dangerously. "There were many turns where the wheels on one side came up off the pavement," Terry says. "We nearly lost that load many times!"

Sixty volunteers, eight years, and many thousands of dollars later, the Best No. 185 was ready to fire up again. That was over 20 years ago and the tractor has been a frequent performer at agricultural events and shows in the central California area, particularly at its home base, Ardenwood Historic Farm.

"Driving the Best is simple," Terry says, "but it takes a lot of people and we do everything quite carefully. We bring the pressure up slowly to 100psi; at about 20psi we open the main steam valve to feed steam to the valve. This tractor incorporates Daniel Best's patented cutoff valve, an invention that lets you go from forward to reverse while the engine is running. The two cylinders are aligned at about 160 degrees, rather than 180, so you never have the problem of getting stuck with the piston at top-dead-center."

Once the steam was brought up, Terry or one of the other authorized drivers blows the whistle. That helps void some of the accumulated water from the steam lines. One single lever on the platform opens all the petcock drains on and around the engine and associated plumbing, draining all the condensate.

"When we are ready to move, we leave the petcocks open, advance the cutoff control *to half-stroke forward*, then we blow the whistle to warn bystanders, then pull the throttle gradually. This puts steam through the system, blows remaining water out, and starts the tractor moving. Once underway we close the petcocks—you normally have to do this with a foot because your hands are occupied—and you're going about 1.5 miles per hour. With everybody aboard, we blow the whistle again, apply maximum tension to the governor, and you're off! It will do about 4.5 miles per hour maximum."

Firing the Best is a full-time job. It is a wood burner and consumes plenty of fuel. The fireman has to control the fire with a practiced eye, keeping in mind the demands of the tractor in the next few minutes. The damper helps cool or heat the fire on the grates. But the fireman also has to maintain water level with the injector or the alternate pump, and keep an eye on the pressure gage.

"Firing the Best is harder than it looks," according to Terry. "When we come to a stop, the steam pressure can build way up. When the safety valve cuts off at 100psi the loud blast will startle the horses nearby and they can bolt. So the fireman needs to know when to douse the fire and reduce the heat to almost nothing *before* we come to one of these halts. The good ones do an excellent job, and the relief valve never goes off." The big tractor has a huge appetite for fuel. Under normal operating conditions, during its working life, the Best No.185 would probably consume about a cord of wood each day.

One of the best fireman on the No. 185 crew is Chuck Wisher. Chuck brings an unusual perspective to the business of steam engines—he's a retired anesthesiologist, professor emeritus of anesthesia at Stanford University. Almost everybody who spends time around a steamer comments on the life-like qualities of steam engines, and Chuck is no exception. "As an anesthesiologist I learned to be always alert to the needs of the patient and the surgeon, and the delivery of anesthetics and non-volatile drugs in surgery isn't automatic. It requires the full attention of the anesthesiologist. I am used to giving my full attention to everything I am doing, and firing a steam tractor requires the same kind of attention."

"Proper firing of a steam engine requires good communication between the engineer and the fireman. The fireman needs to know what to expect and the engineer needs to tell him. Now, emergencies will always occur and sometimes the tractor will have to stop when you're not ready, with a hot fire, and the safety valve will go off—that can't always be helped. But most of the time you can predict what you'll be doing and you must keep things in balance. You must balance the amount of fire on the grate against the amount of steam being consumed. Juggling those factors, plus the amount of air admitted to the firebox and the cooling effect of water being injected, make for the challenge of firing a steam tractor. It takes a great deal of concentration—you can't be distracted by conversation or the passing scene. You have to focus on where you've been, where you want to be, and how you're going to get there. I consider it a personal failure if the safety valve goes off when I am firing."

Daniel Best liked to have a back-up for nearly every critical system on his engines. One of the most important requirements for a steam engine is feed water and No.185 has both an injector, which allows a slow and steady input, plus a big "duplex" pump in the cab that can pump water into the boiler at a much faster rate.

Dan Best may have built a lot of protection into his tractors, but he left out one little thing . . . brakes. You control speed downhill by putting the cutoff in reverse and applying pressure as required to keep speed under control. But if the gears sheared off, and they did, you were strictly out of luck. One of the Best tractors experienced this with a couple of big loaded log wagons in trail astern. The wagons' shoe brakes were applied by the loggers aboard the load, saving the tractor. No. 185 also had a runaway, but broke the frame in the crash.

Avery Undermount Steam Traction Engines Profile: Rick Halldorson's Avery 40 Horsepower

When the steam tractors migrate to Rollag, Minnesota, every year, one machine stands out from the crowd. It is especially tall, really long, and very beautiful—in a muscular, sheet-iron way. This is Jim Briden's 1911 Avery 40 horsepower Traction Engine No. 4249, rescued from the wilds of Alberta, Canada, and brought back to life.

The tractor was built at the very peak of steam tractor popularity, just about the time that gasoline tractors were starting to attract attention. The Avery was shipped north by rail to the little community of Rocky Mountain House, and employed in a sawmill for all its working life. Finally, the woods gave out, the sawmill closed down, and the Avery, along with the rest of the machinery, was left to the elements. The soil in that country is so soft and boggy that only in winter, when the ground is frozen solid, can a heavy load like the Avery move very far.

But in 1969, the tractor was pulled from the wilderness and sold at auction at Kamloops, Alberta. It went through a succession of owners before any attempt was made to restore it. By the late 1970s, a little work had been done on the tractor, some flues were replaced, but it was still essentially a hulk. The cab was gone, the wheels were missing, and it didn't look

It's a long way to the top if ya wanna rock and roll in an Avery undermount steam traction engine. Avery was a Peoria, Illinois, company that offered sturdy, powerful, innovative steam engine designs from the 1890s to the mid 1920s. The company was particularly proud of the undermounted engine placement used on many models.

Steve Gerrode takes Stan Mayberry's Advance Rumley out for a spin. Handling is surprisingly easy considering the vast bulk and weight of these machines. Even with crude steering systems, most will turn on a dime—a *big* dime—and with a dozen turns from lock to lock, perhaps.

much like the glittering iron monster that left Peoria, Illinois, early in the century.

By the late 1970s, a serious campaign had begun to resurrect the Avery. Jim Briden and several other men started searching for parts to replace the multitude of missing components. A cab was fabricated and parts found all over the U.S. for the machine.

The Avery 40 horsepower uses a different design and layout from most of the other tractors shown in this book. The boiler is even heavier than usual, with four rows of rivets; the design is sometimes called the "Alberta Special" perhaps because of the popularity of this style in the Canadian prairie provinces.

"It has been approved to work with 200 pounds of steam pressure," according to Rick Halldorson, one of the Avery's crew. "That is a lot of pressure for an old steam engine, but a testament to the condition of this boiler. Here in Minnesota a boiler must be inspected and rated with a safety factor. The state has certified us to work at that pressure."

Most of the other tractors you see in this book have side-mounted engines, up on the boiler. That makes them somewhat difficult to work on, and hot work, too! The Avery has two cylinders, mounted low; the company called it an "undermounted" configuration. That makes it very easy to do all the routine servicing, such as filling the grease cups and oilers. It also makes it a bit easier for dirt to get into the engine, a problem with plowing. It also puts the boiler rather far off the ground, raising the center of gravity and making the tractor a bit more tippy than others of the same power.

"The two cylinders make for smoother application of power," says Rick, "and they are a lot easier to work on. You don't have to lean up against a hot boiler while you are trying to lube the fittings.

"They call this a 40 horsepower machine, but we've tested it on a prony brake up to 162 horsepower on the belt—that is an actual 'brake horsepower' measurement. The horsepower ratings on these engines were measured many different ways and they don't really tell you a lot. For example, you'll see 15 horsepower steam tractors that can handle maybe seven or eight plow bottoms in average soil; you sure can't do that with a 15 horsepower lawn tractor! So the ratings aren't too informative. Each manufacturer had their own ideas about how they rated their engines.

"From a dead cold start, you can have steam up in about two and a half hours," according to Rick. "It helps to have the wind from behind, into the firebox, and that helps get everything warmed up. That kind of planning was part of being a steam engineer in the past, and still is today. But you don't want to heat up too fast, either, particularly with these old engines. It is a good idea to let them expand and come to life slowly—it's better for the machine overall. They aren't building any more of these things, so we need to take a little extra care of the ones we've got.

"Driving this tractor is a kick. It is enormously large and visibility is a problem. The boiler keeps you from seeing what's going on over on the left, so it is good to have a fireman watching that side and keeping you posted on developments. During the shows, especially, it is easy for a spectator to wander into our path and I can't see them on the left if I am driving."

The Avery uses an excellent steering system they called the *No Slack Positive Screw Shaft Steering Device.* Most other steam tractors use a chain to apply force to the axle, a very sloppy system that is difficult to control. The Avery, on the other hand, is as firm and positive as a British sports car—well, a very heavy, somewhat worn sports car, perhaps, but pretty tight anyway.

Firing the Avery 40 Horsepower

The Avery needs a little special attention from the fireman. Most other tractors can have all the intake air cut off from the fire by shutting the dampers, but the Avery won't let you do that. Instead, the fireman has to anticipate what's likely to happen in the future—will it be more steam, or less?

"You've got to think way ahead," Rick Halldorson says, "because anything you do won't make much difference for maybe five minutes. You can't get excited if you get into a situation where you're under pressured and under powered, that's when you need to just sit in one spot for a while until the pressure builds back up again. Conversely, if you're overfired and high on water, you need to stop and take care of that, too. But you should never get into that kind of situation.

"I've talked to people who've been around an engine when the fusible plug blew. It actually blew out the fire. The fireman had been off the platform when it happened, and would have been severely scalded had he been in his usual spot. The steam blasted out the firebox door and hit the fire with such force that the sheet metal under the grates was blown away and a large crater formed under the tractor."

Profile: Stan Mayberry's Advance

"My folks used to take me to the steam show at Hamilton, Missouri, when I was little, and I was always fascinated by it," Stan reports. "Then Frankie Van Duzen, near here, had a 'thrashing' bee every year; Paul Bright let me play with the engine and run it a little, after they unbelted from the thrashing machine—boy, I got hooked then! I was about 12 years old and started getting involved with the steam tractors in a large way. My uncle, Willie Sykes, had been an old-time steam tractor man and 'thrasherman,' and he'd loan me steam magazines to read. I enjoyed being around those old gentlemen and hearing them tell stories."

Stan bought his first steam tractor shortly thereafter, in 1975, an Advance Rumley that had been used in the Pleasant Hill, Missouri, area. Stan was only 15 years old. The tractor had a long working life, retiring after a few years powering a sawmill. Then it sat outside from the late 1950s until sold at auction in 1975.

It was built in 1919, serial number 15012, probably from the big dealership in Kansas City, Missouri. It is a simple, single cylinder engine with a 9-inch bore and 11-inch stroke rated at 250rpm. Working pressure is 150 or 160 pounds, and with full pressure and an open throttle, Stan can blaze down the road at a head-snapping 2.51 miles per hour. "It weighs about 14 tons," Stan reports, "and the horsepower is supposed to be about 22 horsepower. That's a little controversy with this model. The same engine and same pressure rating has been quoted at both 20 horsepower and 22 horsepower, depending on who's talking."

"On April Fool's Day of 1975 we got it home and went to working on it. We replaced the bottom of the smokebox and the bottom of the flue sheet, a lot of plumbing, the coal bunker, the pipes on the canopy stand, the platform, several flues, and it was ready to go by August."

Stan's magnificent steamer was carefully housed in a machine shed, out of the weather but not out of danger. The shed caught fire in 1984 with the Advance inside. The babbit melted right out of the bearings, the paint came off, and a lot of the plumbing was damaged. Stan spent a couple of winters restoring the steamer all over again.

Stan uses the steamer to power a sawmill on his place from time to time. "The neighbors three or four miles away will sometimes say 'I heard your whistle the other day.' It carries a long ways early in the morning, or late in the evening, when the wind dies down. They have a wonderful sound to them. They aren't too fast, but they are powerful! They don't cost much to run if you use wood or slash from the sawmill. So that's how I got bit by the bug!"

As with any steam tractor, Stan's Advance takes some preparation and planning before firing up. "It takes 450 gallons of water to fill the boiler," he says. "And you need clean water, preferably well water because our city water seems to have too many additives that adds a lot of gunk to the boiler. You can't have any soap such impurities—if you do, it will foam."

With the boiler full of clean, pure water, Stan builds his fire. "It takes me about an hour and a half to get up steam," Stan reports, "although my friend's Advance is steamed up in about an hour. That's from a cold boiler; if you've worked the day before, the boiler will still be warm the next morning and it will heat up faster. Use a lot of kindling—and be as sloppy as you can when you're building your fire. That means with the wood every which way on the grate so lots of air can get through."

When you fire a boiler with wood you keep the firebox pretty much full all the time. Coal is a different proposition, and too much will slow air flow and block the heat. "You want to open your fire door just long enough to chuck in a couple of pieces, then close the door to keep the cold air out. With coal, you have to watch the corners—they burn out first; keep a thin fire, fire often in small quantities. Black smoke from the stack means wasted coal!"

It is quite important to keep the ashes under control. If they accumulate to the point where they pile up and touch the grates, warping is likely to destroy the grate. The fireman or engineer will clean out the ash pan as often as three times a day.

APPENDIX

The following is from "Kernels for the Starved Rooster," an occasional publication of the Aultman & Taylor Machinery Company, 1902.

Defects

Some of the more prominent defects of boilers are here mentioned, together with their causes and the usual remedies therefor:

Loose Rivets.–This is generally the effect of overheating. They should be cut out and new ones driven; but in case the rivets are too small for the thickness of the plates, and especially in a girth seam, the rivets must be replaced by others of a larger diameter.

Blisters.–These are due to imperfect welding in the manufacture of the plates. They should be trimmed off to ascertain their extent and thickness, and, if of small area and slight thickness are not dangerous; but if the contrary is the case, or if the plate is cracked under the blister, they must be cut out, and a "hard patch " put on inside the boiler to avoid making a pocket for the collection of sediment. Sometimes this patch will be found to leak, and caulking will not stop it, in which case, on removing the patch previous to renewal it will generally be found that there are cracks under the heads of the rivets.

Burnt Plates.–These are due to "low water," to a deposit of sediment or scale, to continued impact of flame caused by leaks of air through the masonry, and when a seam is just back of a bridge wall; but sometimes it is caused from an encrustation, or soap formed from oily matter. The place should be cut out and a hard patch put on.

Bagging, Buckling or Bulging of plates, sometimes forming a pocket, generally occurs from the overheating of the iron in consequence of deposits of oil, sediment, or scale. Sometimes it occurs when the boilers are clean, and then it is the result of impact of flame, or it may be caused by the unequal expansion of the various laminae of the plates under their daily usage in case their are no signs of overheating. It is very liable to occur in the flat fire-sheets of the sides of fire-box boilers when the stays are spaced too far apart. The usual remedy in this case is to put in an additional stay-bolt between each four stays already there. The presence of a bulge on the bottom fire-sheet of a boiler is not necessarily dangerous, but it must be watched carefully, and its surface kept clean, and, at the first indication of weakness, it must be heated and forced back into place, or else cut out and a hard patch put on.

Steam Leaks are generally considered fair evidence of overheating, especially in the lower half of the girth seams. They should be carefully examined, the rivets having been cut out, and if any evidences of cracks under the heads of the rivets are found, the seam should be cut out also and a patch put on. Sometimes a leak occurs in a seam from the rivets being too small, or from the lap being too great. In the former case the rivets should be replaced by larger ones, and in the latter the seam must be chipped and caulked.

Westinghouse Threshing Machinery

Cracks in plates may be due to overheating, or to unequal contraction and expansion, or to letting cold water strike on a hot plate, and this last often happens in hurrying to clean a boiler.

As a general rule, it is safe to say that accidents from the overheating of boiler surfaces do not occur at the moment of overheating, but at some subsequent period. How soon depends upon the extent of the overheating. Steam boilers can be used with almost perfect security if proper attention is paid to them, requiring merely a careful observance of natural laws, and the constant exertion of that much-neglected faculty called common sense.

Neglect of the masonry in the setting of a boiler is often the cause of external corrosion, and cracks or loose bricks should never be allowed. Nothing but fire-clay or kaolin should be used to cement the bricks wherever they come in contact with the boiler.

The braces of a boiler need careful attention. All scale and rust should be removed from them, and they should be sounded with smart blows of the hammer at the extremities to detect slacks or breaks. A broken brace should always be repaired as soon as discovered. The gauge-cocks, feed and blow-cocks, when found to leak, should be ground in as shortly thereafter as possible; and, if they are choked, the obstructions must be removed. It is well to open the blow-cock a little once every day to prevent its setting fast. If a back stave is broken, a new one must be put in its place at once, or the walls may begin to bulge or crack.

The cast-iron front often cracks, and it should be repaired by bolting to it a piece of Iron plate or bar to prevent widening of the crack or sagging, unless a new section can be readily procured and substituted for it, which is preferable.

Leaky feed and suction pipes must be repaired by parceling and serving, if of copper; but if the pipe is of iron, it should be thrown aside and a new length substituted.

If a gauge-glass breaks, shut off the water first, and the steam afterwards, to avoid being scalded.

If a bolt blows out or a tube splits, drive in a pine plug till you have time to repair it properly, which should be after work is over for that day.

The brick lining of the furnace is always getting out of order, especially at the front and at the bridges; don't wait too long before you make the necessary repairs. Remember that prompt repairs save long bills.

Explosions

Explosions of steam boilers are due to defective material, natural deterioration, defects of construction, or to improper management; they take place when the resistance is less at some point than the pressure to which it is subjected, and may happen even when the pressure is very low.

The steam pressure in a boiler is never increased otherwise than by the natural cumulative effect of the furnace. There exists no phenomenon that has power to suddenly increase the power in a boiler. The action which has been attributed in this respect to the " spherical state" and to the " superheating of the water" has not been confirmed by observation.

The explosion of a boiler is not an instantaneous action, though seemingly so; it is a well-defined and rapidly succeeding series of operations. The rupture commences at the point where the resistance offered by the material is less than the strain to which it is subjected, and it extends into the adjoining parts when these parts are too weak to sustain this increased strain that the rupture already made brings to bear on them, together with the shock due to the motion that the edges of the fracture make while seeking a new state of equilibrium.

The number and direction of the ruptures depend especially upon the resistance of the parts adjacent to the first fracture. A rupture, even when of considerable extent, does not produce an explosion if the adjoining parts possess sufficient strength.

In case of an explosion the steam pressure does not fall the instant that the rupture takes place. On the contrary, the pressure continues very nearly constant up to the time when all the water has escaped from the boiler

An explosion is so much the more terrible as there are more fractures made prior to the moment when the boiler is entirely emptied of its water.

It is very dangerous to let the water get so low in a boiler that the plates become red-hot, because the softened plate will tear open, and may produce an explosion if the red-hot part is of large extent, or if the adjoining parts do not offer sufficient resistance.

When water is fed into a boiler where the water is too low, it almost invariably lowers the pressure of the steam. So that it is always dangerous to introduce feed before having dampened the fires with wetted small coal and ashes, or having drawn them, because the water injected quiets the ebulition, and increases the surface exposed to heat.

There is not, nor can there be, any connection between the explosion of steam boilers and the phenomenon known as " the spheroidal state."

Even when "low water " does not cause the plates to be heated to redness, it causes leaks in the seams and at the tube ends, and fractures also by reason of the unequal contraction and expansion produced.

It is dangerous to empty a boiler when the flues or tubes are still hot, or to fill a boiler with cold water before it becomes sufficiently cool, or to wash it out while it is still warm, for such action causes fractures of the transverse riveting in such a manner as may not always be shown by leakage, and this defect may very easily produce an explosion when next the fires are lighted, or in a short time afterward.

It is also dangerous to fire up under a boiler too rapidly, as when the draft and combustion are sufficient for a " white heat," the plates, no matter how good they are, can not resist with certainty.

Explosions are generally due to unseen defects, because no one has tried to discover them.

Overheating of a boiler may also be caused by accumulation of scale or sediment, or rather foreign matter, on the furnace plates, flues, or tubes and heads, or by the metal being too thick near the fire, and by defective circulation.

Defective circulation is generally due to the design of the boiler, from its having too cramped water spaces; from the connection being impeded by the overcrowding of tubes, or placing them too close over the furnace crowns; and from having too large a body of dead water lying below the heating surfaces, which causes unequal contraction and expansion.

Fairbairn states that the efficiency and safety of a boiler depend as much upon the efficacy of the circulation as they do upon the strength and disposal of the material of which it is composed.

The hydraulic pressure applied to a boiler does not show us whether it is dangerous to use it or not. Many boilers have exploded when corrosion has reduced the metal so much that a smart blow from a ball-faced hammer would have knocked a hole in the sheet, while the boiler has given no evidence of weakness under the hydraulic pressure.

Another source of explosions is due to the sudden and unequal expansion of the metal of large cast-iron stop-valve chambers when steam is suddenly let on in cold, frosty weather through opening the valve, water of condensation being present. Such explosions have taken place with only ten pounds pressure.

Boilers can also explode from a preliminary explosion of gas in the furnace or flues.

A thorough internal and external inspection of boilers by a person skilled in the profession is the only means of ascertaining their condition. Each and every part of a boiler may contain dangerous defects, and an examination is only finished when every part of the boiler has been carefully inspected.

When a plate is covered with soot or encrustation most of the defects can not be seen; therefore it is highly important that boilers should be kept as clean as possible, as well externally as internally.

It is a bad plan to hasten the cooling of a boiler, especially if it is a long one, and it is very injurious, also, to let the water out of a boiler until the tubes or flues are sufficiently cooled. Experience has shown that out of every ten accidents or explosions that happen only one occurs to a regularly inspected boiler. From what has been said it is plain that a boiler may burst and not explode, but that an explosion is always preceded by a bursting to which the explosion is a consequent.

A Bad Rupture

The various steps in an explosion, which have been before mentioned, may be defined as follows:

1st–A fracture in a plate followed by a rending.

2nd–A violent outburst of water and steam.

3rd–A fall in pressure.

4th–Portions of water are propelled with great violence against the shell of the boiler, and shattering it by the expansive force of the steam disseminated throughout the body of the water.

5th–The steam generated from the liberated water imparts a high velocity to the fragments, converting them into projectiles, and thus spreading ruin and destruction around; and also widely scattering the particles of water not converted into steam.

It is generally said, by those not thoroughly conversant with steam, that, in case of an explosion, the water must have been low; and they point in corroboration of their statement, to the fact that no water is found scattered round, Now nothing is more certain than that all of the water contained in the boiler can not be converted into steam when an explosion occurs, for 965 units of the heat are required to boil water into steam from and at 2120 Fahrenheit: while at 140 pounds pressure the temperature would be 3610, or 149 units higher than 2120. Now, 965 divided by 149 is 6 1/2 nearly. That is to say, of every 6 1/2 Lbs. of water in the boiler one pound only will be converted into steam of atmospheric pressure, while 5.5 Lbs. of water will be scattered in the air, mixed with the escaping steam.

From various experiments and investigations the following conclusions have been arrived at:

A violent explosion may take place in a boiler when there is plenty of water in it.

A moderate pressure of steam may produce a terrific explosion when there is plenty of water.

The American Thresherman

That a boiler may explode under steam at a less pressure than it has stood without apparent injury from a water pressure.

A rupture will be followed by relief of pressure, with or without explosion, as the fracture is intended or otherwise.

That an explosion rarely occurs in an externally fired boiler from "low water."

In examining into the cause of a boiler explosion it is generally necessary to determine the initial point of rupture, which often will solve the whole question; but before making a final decision we should be prepared to show:

That the cause can exist in the case in question.

That it is competent to produce the results ascribed to it.

That no other known cause can produce these effects.

Fuel.

There is often required, for the use of the boilers of the average manufactory, nearly twice as much fuel as would be necessary if the plant were of a better type and more skillfully managed; and in some places the waste is greater than this.

The way in which this waste takes place may be stated generally as follows:

The engines require too much steam to develop the power required.

The boilers are badly designed, improperly set, or too small.

The coal used is of poor quality, or improperly housed.

The firemen are careless or ignorant of their business.

The proper course for an owner to pursue, in case he suspects waste, is to employ a skilled expert engineer to examine his plant thoroughly and report upon the defects discovered, stating the remedies therefor; and then act upon the report as he would in any other matter of business.

Experiments have shown that coal loses from ten to forty per centum of its evaporative effect from being exposed to the weather. It should always be kept under cover, and the building should be of brick and closed in.

It is generally conceded that 2 Lbs. of good dry wood are equivalent in evaporative effect to one pound of good coal; but it must be remembered that wood requires a roomier furnace than coal, and also that the spaces between the grate bars must be larger.

The fuel value of the same weight of different woods is very nearly the same–that is a pound of hickory is worth no more for fuel than a pound of pine, assuming both to be dry.

If the value be measured by weight, it is important that the wood be dry, as each ten per centum of water or moisture in the wood will detract about twelve per centum from its value as a fuel.

The weights of one cord of different woods (air dried), as well as the fuel value in comparison with coal, is as follows:

1 cord hickory or hard maple weighs 4,500 Lbs. and is equivalent to 2,000 Lbs. of coal.

1 cord white oak weighs 3,850 Lbs. and is equivalent to 1,715 Lbs, of coal.

1 cord beech, red oak, black oak weighs 3,250 Lbs, and is equivalent to 1,450 Lbs, of coal.

1 cord poplar (white wood), chestnut, elm weighs 2,350 Lbs. and is equivalent to 1,050 Lbs. of coal.

1 cord the average of pine weighs 2,000 Lbs. and is equivalent to 925 Lbs. of coal.

"Science of Successful Threshing" (1899) Excerpts

The Case Company sent a little handbook along with steam traction engines and threshing machines called *The Science of Successful Threshing*. This little volume provided a guide for the new owner on how to set up and operate the new machine. Here are some of the helpful hints in the book:

HILLS AND MUD HOLES

In coming to a steep hill the engineer should first see that he has about the right amount of water in the boiler, that is, about 2 inches in the glass with the boiler level. With the boiler too full there may be danger of priming which should be especially avoided on a hill.

With too little water the crown-sheet may be bared going down the other side. It is necessary to exercise a little judgment also in regard to the fire. It should he hot enough to insure steam to go up the hill without stopping but don't let the engine blow when she is pulling hard on a hill, for this is liable to cause priming, which would necessitate stopping. In short, when approaching a steep hill prepare for it so that you know you can go up and down without stopping.

Every man in charge of a traction engine with a locomotive boiler must know the danger of stopping with the front end low In ascending a hill don't try to see how fast you can go, but run at a moderate speed.

In going up steep gravel hills there is danger of breaking through the surface crust, letting the traction wheels into the soft gravel, which they will push out from under them, simply digging holes instead of propelling the engine. When this occurs, don't let the engine bury itself, but stop at once. Block the wheels of the separator, or whatever the engine is pulling, and uncouple the engine and the chances are that it will go out all right. If it does not, put cordwood sticks in front of the traction wheels so the grouters will catch or else hitch on a team and start the team and engine together. In going down a very steep hill leave the throttle partly open to admit a little steam and if the engine runs too fast control the speed with the reverse lever.

What has been said about gravel hills applies in general to soft mud holes. Stop the engine when the wheels slip and put straw, brush stones, sticks or anything else that may be handy in front of the wheels. When the engine is on a "greasy" road where the wheels slip without digging much, get a couple of men to roll the front wheels and you will be surprised how much good they will do. With one traction wheel in a greasy mud hole or old stack bottom and the other on solid ground the differential gear may be locked, but unless you understand the consequences of doing this it will be better to get out some other way.

It is a good plan to carry a heavy rope with the machine. Then when the engine gets "stalled" it can be uncoupled and run onto solid ground where it can pull the separator out of the hole by the rope and then be coupled up short again. A rope is elastic and therefore better than a chain, which is liable to snap with the shock of starting the load. The rope should have a ring spliced in one end and the other may be tied into a clevis on the engine draw-bar in a "bow-line" knot, which will not slip and is easily untied after being strained. If a chain is used the engine must be moved very slowly until all the slack is out of the chain.

Engines for Louisiana and other swampy countries have wrought iron "mud-claws" to bolt to the traction wheels in addition to the regular grouters. They are about 5 inches high and must be taken off before crossing bridges.

Cleaning the Boiler

No rule can be given as to how often a boiler should be cleaned. In some localities it will be necessary to clean it twice a week, while in others, where the water is almost perfectly clean and pure, once in six weeks will be often enough. To clean the boiler remove the plugs or `hand-hole' plates in the water-leg and the brass plug at bottom of front tube-sheet. Wash the boiler thoroughly with a hose. In replacing hand-hole plates be sure the packing is straight and that the inside of sheet has been cleaned of bits of scale that may have adhered to it. Screw up the nuts moderately tight and tighten them a little more when steam begins to show on the gauge.

In emptying the boiler previous to cleaning, make sure all of the fire is out and the steam below 10 pounds pressure before opening the blow-off valve. This is necessary, to prevent the mud from becoming baked on the tubes and sheets.

LUBRICATION

Keep your engine well oiled if you would have it last and not cause trouble. By well oiled is not meant that it should be swimming in oil, but that the bearings be always lubricated. It does not take very much good oil to keep a bearing properly lubricated, but you should apply it often and be sure that it reaches the place intended.

Use a good quality of valve Oil in the lubricator, as it is very important that the packing rings and valve should be well lubricated with an oil that will stand the high temperatures of the steam. Keep the lubricator feeding every minute the engine runs.

Hard Oil has many qualities to recommend it. It stays on the bearing, and as it wears well a little of it will go a long way.

Buy your oils of someone that can he depended upon to give you a good article. Many of the oils now market are largely adulterated with rosin and paraffin, and though having the appearance of an excellent oil, have poor lubricating qualities, are gummy and dry up in a short time.

NOTES

The appearance of an engine reflects the character of the engineer. Keep your engine clean.

An engine is said to be "'running over" when the top of rim of fly-wheel runs away from the cylinder and "running under" when the top of rim of fly-wheel runs towards the cylinder.

The smoke-box and stack and occasionally the whole boiler will require blacking. For this use asphaltum or boiled linseed-oil with a little lamp-black in it.

An engine is on its "dead-center" when a line drawn through the center of the piston-rod will pass through the center of the crank-pin. There are two, the "crank" dead-center, when the piston is at the end of the cylinder nearest the crank-shaft, and the "head" dead-center, when the piston is at the opposite end.

In cold weather be sure that all pipes and fittings liable to freeze have been drained before leaving the engine for the night. It is a good plan to cover the engine with a canvas to keep in the heat.

Upon returning to the engine after an absence always open the pet-cock under water-gauge and see the water move and return to the same height or else try the gauge cocks. In either case be sure you know where the water is and not just think you know.

There is rarely an excuse for keeping a threshing crew waiting for an engineer or his engine. He should always be ready to start (or stop) the instant the signal is given from the separator.

Warped and burnt grates show that the ashes have been allowed to fill up and exclude the air. A good engineer never lets this occur.

Steam Shows Around America

Pioneer Peanut Days, presented by the Dixie Flywheelers Association, Post Office Box 6363, Dothan, Alabama, 36302, normally toward the end of October.

California Antique Farm Show, presented by Early Days Gas Engine & Tractor Club Branch 8, Post Office Box 1475, Tulare, California, 93275, normally in late April.

Semi-Annual Threshing Bee, Antique Tractor & Engine Show, Early Days Gas Engine & Tractor Assn., 2040 North Santa Fe Road, Vista, California, 92024, normally in mid-June.

Connecticut Antique Tractor Show and Pull, Route. 169, Brooklyn Fairgrounds, Brooklyn, Connecticut. For info call 1-800-442-5182. Normally held in mid-October.

Sunbelt Ag Expo, Post Office Box 28, Tifton, Georgia, 31793, but held at the fairgrounds at Moultrie in mid-October.

River Valley Antique Association's Threshing Show and Antique Display, late September, Three Sisters Park south of Chillicothe, Illinois, on Route. 29.

Fall Festival, Boonville, Indiana, held in mid-October. Info: (812) 925-7666.

And a whole lot more can be found in the Antique Power Show Guide, Post Office Box 562, Yellow Springs, Ohio, 45387; 1-800-767-5828.

INDEX